Creating and Managing Environments for Telecommunication Systems

Creating and Managing Environments for Telecommunication Systems

Edited by **Claude McMillan**

Willford Press
New York

Published by Willford Press,
118-35 Queens Blvd., Suite 400,
Forest Hills, NY 11375, USA
www.willfordpress.com

Creating and Managing Environments for Telecommunication Systems
Edited by Claude McMillan

International Standard Book Number: 978-1-68285-165-4 (Hardback)

Printed in China.

Contents

Preface

The purpose of the book is to provide a glimpse into the dynamics and to present opinions and studies of some of the scientists engaged in the development of new ideas in the field from very different standpoints. This book will prove useful to students and researchers owing to its high content quality.

Telecommunication systems have revolutionized the way we communicated in the past decades. With the progress in technology, these systems have also continuously evolved. Analysis and performance, optimization strategies, wireless sensor networks, etc. are some of the areas that have been delved into in a lucid manner, in this book. It aims to equip students and experts with the advanced topics and upcoming concepts in the field of telecommunications. The insightful contributions of an internationally renowned panel of scholars make this book apt for students and professionals.

At the end, I would like to appreciate all the efforts made by the authors in completing their chapters professionally. I express my deepest gratitude to all of them for contributing to this book by sharing their valuable works. A special thanks to my family and friends for their constant support in this journey.

Editor

1

Data Confidentiality in Mobile Ad hoc Networks

Hamza Aldabbas, Tariq Alwada'n, Helge Janicke, Ali Al-Bayatti

Software Technology Research Laboratory (STRL), De Montfort University, Leicester, United Kingdom

{hamza, tariq, heljanic, alihmohd}@dmu.ac.uk

ABSTRACT

Mobile ad hoc networks (MANETs) are self-configuring infrastructure-less networks comprised of mobile nodes that communicate over wireless links without any central control on a peer-to-peer basis. These individual nodes act as routers to forward both their own data and also their neighbours' data by sending and receiving packets to and from other nodes in the network. The relatively easy configuration and the quick deployment make ad hoc networks suitable the emergency situations (such as human or natural disasters) and for military units in enemy territory. Securing data dissemination between these nodes in such networks, however, is a very challenging task. Exposing such information to anyone else other than the intended nodes could cause a privacy and confidentiality breach, particularly in military scenarios. In this paper we present a novel framework to enhance the privacy and data confidentiality in mobile ad hoc networks by attaching the originator policies to the messages as they are sent between nodes. We evaluate our framework using the Network Simulator (NS-2) to check whether the privacy and confidentiality of the originator are met. For this we implemented the Policy Enforcement Points (PEPs), as NS-2 agents that manage and enforce the policies attached to packets at every node in the MANET.

KEYWORDS

MANETs, Policy Enforcement Point(PEP), Policy decision Point(PDP) and Discretionary Access Control (DAC).

1. INTRODUCTION

Mobile *ad hoc* networks are autonomous systems comprised of a number of mobile nodes that communicate using wireless transmission. They are self-organized, self-configured and self-controlled infrastructure-less networks. This kind of network has the advantage of being able to be set up and deployed quickly because it has a simple infrastructure set-up and no central administration [1]. Obvious examples are in the military or the emergency services. One scenario is establishing communication between various agents in a disaster recovery operation where e.g. fire fighters need to connect to local ambulances and traffic control in circumstances where the normal communication infrastructure is destroyed or otherwise rendered unusable. In such situations a collection of mobile nodes with wireless network interface can form a transitory network [2]. These networks are particularly useful to those mobile users who need to communicate in situations where no fixed wired infrastructures are available. However, the salient feature of creating a network 'on the fly' without requiring any prearranged infrastructure gave mobile *ad hoc* networks an appreciated interest in both industrial and military systems. The key challenges in MANETs design come from the decentralised nature, self-organisation, self-management, and also the fact that all communications are carried over wireless links in short-range communication [3]. In addition to that the topology in the network is dynamically changed because of the high mobility nature in (MANET). Therefore, all these unique characteristics present appreciable challenges for MANETs [4, 5, 6].

In comparison with wired networks where the devices must have physical access to the network medium, mobile *ad hoc* networks have no apparent secure boundary. Once the attackers are in the transmission range of any other device, then they can join and communicate with all devices in range. Indeed the nature of mobility in *ad hoc* networks, and the liberty to join, move and leave networks makes the MANET vulnerable to attacks, which can result from any device in the same transmission range [3].

Whilst devices used in wired networks get their electrical supply directly through available power grids, in MANET nodes are generally operated by small batteries with a limited lifetime. This makes nodes unable to perform intensive computations over prolonged periods of time. An attacker on the other hand is typically able to provide sufficient power-supply and thus must be assumed to be able to perform intensive computations [1], meaning that attack and defence in these networks is not levelled. The lack of centralized management in MANET makes the detection of attacks a very complicated issue. Mobile *ad hoc* networks are highly dynamic and large scale, and they cannot be easily monitored; benign (non-malignant) failures in MANETs are fairly common, e.g. transmission destructions and packet dropping. As a result, malicious failures will be more difficult to identify. Since security is an essential component in a hostile environment, these unique characteristics of mobile *ad hoc* networks raise challenges that security requirements must address [7, 8].

There has been appreciable work by the research community [7, 9, 10, 11, 12] in message encryption, digital signature, key management, etc. that address the confidentiality of messages in transit over the unsecured medium. Many challenges particularly related to the privacy and data confidentiality, however, remain to be solved. Existing approaches which have been used in MANETs such as access control, digital signature, and encryption focused only in securing the channel during the transmission, however how these nodes act after and use this information has been mostly neglected.

In this paper, we review the main security issues and existing solutions in MANET, in particular the area of security of MANET which has not been widely addressed. We present a policy-based architecture to control the data dissemination in MANETs that is build on the automatic communication of policies between nodes and draws on concepts developed for other forms of networks e.g. [28]. The purpose of this architecture is to keep data confidential not only during communication, but also after it has been transmitted to another node. Ensuring that the contents of messages are kept secret to an originator-defined subset of peers in MANETs.

In this paper, we highlight the characteristics in MANETs in Section 2, and focus on security issues in Section 3. In Section 4 we present some of the previous work on securing MANETs to which we relate our proposed policy-based architecture and the algorithm for updating policies in Section 5. We discuss advantages and limitations of our approach in section 6 and conclude the paper in Section 7 where we also summarise our findings and outline our future work in this area.

2. Characteristics of MANET:

Mobile *ad hoc* networks (MANET) are autonomous systems of mobile nodes connected by wireless links. These nodes are therefore free to move arbitrarily; thus, the topology of wireless networks can be changed dynamically and unpredictably. MANETs have many characteristics that make them are different from other wireless and wired networks that are well recognised [13, 14, 15, 16, 8]:

1. **Multi-hop communications**: The communication in MANET between any two remote nodes is performed by numerous intermediary nodes whose functions are to relay data-packets from one point to another. Thus, *ad hoc* network requires the support of multi-hop communications. For example, in Figure 1, nodes A and D must engage the help of nodes B and C to relay data-packets between them in order to communicate.

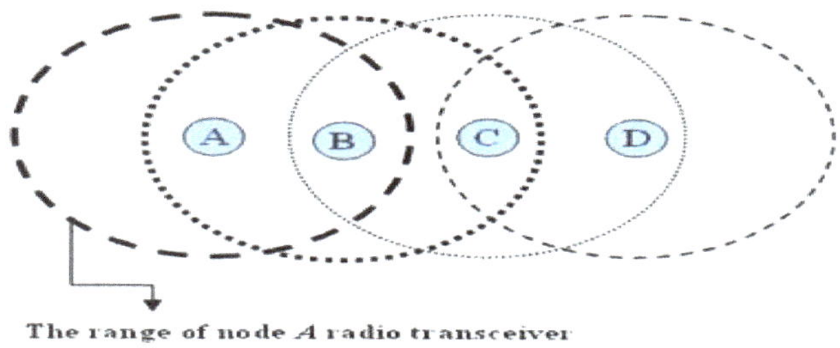

Figure 1: MANET of four nodes and their transmission ranges

2. **Constrained Resources:** Generally, most MANET devices are small hand-held devices ranging from personal digital assistants (PDAs) and laptops down to cell phones. These devices indeed have limitations because of their restricted nature; they are often battery-operated, with small processing and storage facilities.
3. **Infrastructure less:** MANETs are formed based on the collaboration between autonomous nodes, peer-to-peer nodes that need to communicate with each other for special purpose, without any pre-planned or base station.
4. **Dynamic Topology:** MANET nodes are free to move, hence the connectivity between nodes in MANET can change with time, because nodes can move arbitrarily; thus the nodes can be dynamically inside and outside the network, constantly changing their links and topology, leading to change in the routing information all the time due to the movement of the nodes. Therefore, the communicated links between nodes in MANET can be bi-directional or unidirectional.
5. **Limited Device Security:** MANETs devices are usually small and can be transported from one place to another, then they are not constrained by location. Unfortunately, as a result these devices could be easily lost, stolen or damaged.
6. **Limited Physical Security:** Generally, MANETs are more susceptible to physical layer's attacks than wired network; the possibility of spoofing, eavesdropping, jamming and denial of service (DoS) attacks should be carefully considered. By contrast the decentralised nature of MANET makes them better protected against single failure points.
7. **Short Range Connectivity:** MANETs rely on radio frequency (RF) technology to connect, which is in general considered to be short range communication. For that reason, the nodes that want to communicate directly need to be in the close frequency range of each other. In order to deal with this limitation, multi-hop routing mechanisms have therefore to be used to connect distant nodes through intermediary ones that operate as routers.

3. Network Security Considerations

Since security is an essential component in MANET, the striking features of mobile *ad hoc* networks raise both challenges and opportunities in achieving these security goals, Unlike other traditional networks (wired) where nodes must have physical access to the network or communicate through several defence perimeters like firewalls and gateways, MANET uses the wireless medium so attacks on a wireless network can come from all directions and target any node. This provides a larger surface of attack ranging from passive attacks, such as "tapping" to active attacks, such as message replay, message leakage, contamination and distortion. This means that a MANET does not have a clear line of defence, and every node must be prepared to defend against the different kind of attacks [17].

MANETs present a new set of challenges to security design as security solutions must achieve both data protection and efficient network performance [18]. While we addressing network security, we have to consider the security requirements to take account of the functionality required to provide a secure networking system.

3.1. Security Requirements

The security requirements specified below specified by International Telecommunications Union (ITU-T) represented in their recommendation X.805 and X.800 [19, 4, 18, 29]:

1. **Authentication:** Authentication is essential to verify the identity of every node in MANET and its eligibility to access the network. This means that, nodes in MANETs are required to verify the identities of the communicated entities in the network, to make sure that these nodes are communicating with the correct entity.

2. **Authorisation and Access Control:** Each node in MANET is required to have the access to shared resources, services and personal information on the network. In addition, nodes should be capable of restricting each other from accessing their private information. There are many techniques that can be used for access control such as Discretionary Access Control (DAC), Mandatory Access Control (MAC) and Role Based Access Control (RBAC).

3. **Privacy and confidentiality:** Each node has to secure both the information that is exchanged between each other; and secure the location information and the data stored on these nodes. Privacy means preventing the identity and the location of the nodes from being disclosed to any other entities, while confidentiality means keeping the secrecy of the exchanged data from being revealed to those who have not permission to access it.

4. **Availability and survivability:** The network services and applications in MANET should be accessible, when needed, even in the presence of faults or malicious attack such as denial-of-service attack (DoS). While survivability means the capability of the network to restore its normal services under such these conditions. These two requirements should be supported in MANET.

5. **Data integrity:** The data transmitted between nodes in MANET should be received to the intended entities without been tampered with or changed by unauthorised modification. This requirement is essential especially in military, banking and aircraft control systems, where data modification would make potential damage.

6. **Non-repudiation:** This ensures that nodes in MANET when sending or receiving data-packets should not be able to deny their responsibilities of those actions. This requirement is essential especially when disputes are investigated to determine the misbehaved entity. Therefore digital signature technique is used to achieve this requirement to prove that the message was received from or sent by the alleged node.

4. State Of The Art

Existing approaches in security which have been applied to MANETs are for example using traditional cryptographic solutions based on public key certificates [24] to maintain trust, in which a Trusted Third Party (TTP) or Certificate Authority (CA) [25] certifies the identity associated with a public key of each communicated entities. Iman Almomani and Hussein Zedan [26] proposed a comprehensive, top-down, end-to-end security solution for MANET based upon a well defined architecture and exploiting two of the ITU-T recommendations: X.800, and X.805. Such approaches can therefore provide end-to-end secure communication channels and are mainly focused on message confidentiality, integrity and non-repudiation, they do not consider however the trust management of the communicated entities, and how these certified entities act is left to the application layer [20]. Ali Al-Bayatti *et al* [27] proposed behaviour detection algorithm combined with threshold cryptography digital certificates to satisfy prevention and detection to securely manage Mobile *Ad hoc* Network of Networks (MANoNs). Lidong Zhou *et al* [10] studied the security threats, vulnerabilities and challenges which faces the *ad hoc* network. In their work [10] they protected the packets sent between nodes by choosing the secure routing path to the destination node based on the redundancies routes between nodes to maintain the availability requirement. This is because all key-based cryptographic approaches such as digital signature need a proper and secure key management scheme to bind between the public and private keys to the nodes in the network; Lidong Zhou used replication and a new cryptographic technique (threshold cryptography) [21, 22] to build a secure key management process to achieve trust between a set of servers in *ad hoc* networks by distributing trust among aggregation of nodes to certify that these nodes are trustworthy.

Securing the routing in mobile *ad hoc* network (MANET) has also been given much attention by the researchers; many approaches, therefore, have been proposed to deal with external attack. Sirios and Kent [23] proposed an approach to protect the packet sent to multi receivers by using keyed one-way hash function supported by windowed sequence number to ensure data integrity. The trust issue in communication systems as in mobile *ad hoc* network is a challenging task to achieve. The framework that we are proposing has some relations with secure routing protocols, as it determines the sharing of information, however we consider policies to operate at a higher level in the protocol stack where application specific trust decisions can be made. Public Key Infrastructure (PKI) and cryptography are achieving a kind of a quasi-trust before the communication is started. However, how the nodes act after that is a controversial issue as untrusted nodes cannot be predicted without establishing tracing techniques to ensure that they are not misbehaving whilst participating in the MANET.

In commercial and medical contexts, individuals should be able to keep and manage access to their personal information by choosing to which entities information should be disclosed in a discretionary way. Individuals expect that their personal data, personal information, or Personally identifiable information (PII) to be securely communicated and shred between different organisations such as their names, addresses, phone numbers, national insurance numbers, credit card details, passwords, or date of birth (DOB). Therefore, Siani Pearson and Marco Casassa Mont [28] employed a clever idea of sticking policies with data to control how the PII should be processed, handled, shared with other parties.

5. Our Proposed Framework

In this paper, we show how revealing secret information by a malicious node (inside the network) to unwanted nodes could cause a privacy and confidentiality breach. Therefore traditional encryption tools being commonly used solve one part of the problem by encrypting data exchanged between nodes using the public key of the destination node and then decrypting the packet by the destination's private key but how the destination behave after is neglected. Using a mechanism which employs access control to ensure confidentiality is a real possibility, so our work intends to use this mechanism, in particular Discretionary Access Control (DAC) to ensure data confidentiality and privacy of the originator node in MANETs.

5.1. Scenario

To motivate our approach we present a scenario in which a close military alliance between two countries wants to share tactical mission information only between themselves and not with other coalition members.

Considering three nodes A,B,C in Figure 2, where node A and B are allocated to British and US armies, and C is allocated to Afghan army. Node A wants to send a tactical message for the mission that says "'we are going to start the mission after 8 pm"' to node B, however node A does not want node B send the message to node C because node C is not trusted by A. How can node A trust node B not to send the message to node C?

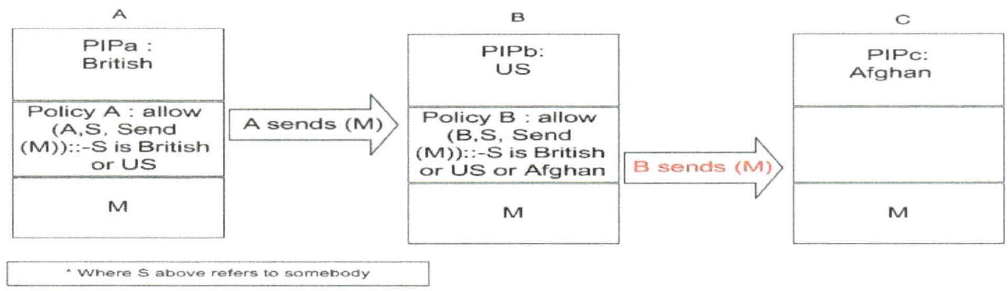

Figure 2: node B disclose the message to C

Node A sends the message (M) to node B, node B now knows the message (M). However depend on its policy node B can send the message (M) and disclose it to node C. Which it is the problem of the node A privacy.

The goal of our proposed approach is to solve this problem by allowing the originator to specify a high level policy which will automatically apply and enforce itself to all the communicated entities on the network. This is done by attaching the policy of the originator (A) with the message (M) to control the access to it, which is capable to define who are allowed to access that message. In this way the policy of node A attached with the message (M), tells node B to which node can the message (M) be send to (only British or US armies can receive the message) as in Figure 3 :

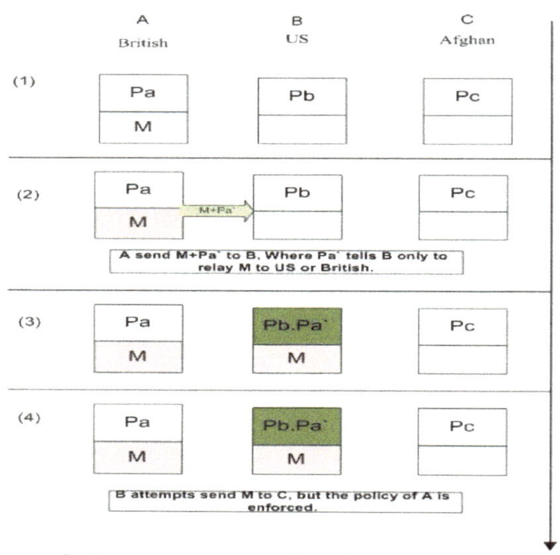

Figure 3: Prevention Node B Of Disclosing M to node C

node A sends the message (M) + policy of node A which tells node B to allow sending the message (M) to any node if its British or US nodes. Here after node A sent the message (M) to node B attached by the node A policy. node B receives the packet and now knows the message (M) in addition of that it knows the policy of A:

Allow (Node B, Send(M) to S: if S is British or US, where S refers to somebody.

5.2 Our Framework

Figure 4 presents the proposed framework, where policies are used to enforce access control to such information sent by the originator to other entities in the system, our framework will be introduced in every entity in the communicated systems.

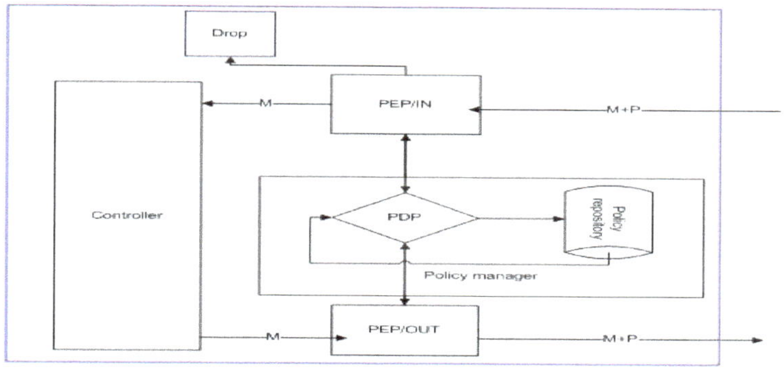

Figure 4: The proposed framework

Our framework is composed of four components as they shown in the Figure 4:

1. Policy Enforcement Point (PEP/OUT): executes and enforces policy decisions in the sender node, this component installed at the transmitter interface that does merge system's policy with the message sent to others systems. In our simulation we configured the send function to achieve the functionalities of this component.
2. Policy Enforcement Point (PEP/IN): executes and enforces policy decisions in the receiver node, this component installed at the receiving interface that does inverse process at the receptive system, splitting and dividing the message from the policy attached. In our simulation we configured the receive function to achieve the functionalities of this component.
3. Policy Decision Point (PDP): plays a crucial role in both the sender and the receiver side in our framework, and helping other components to do their jobs. In our simulation we configured this function as a shared source code to achieve the functionalities of this component in all nodes.
4. Controller that process and store the information received from the other components.

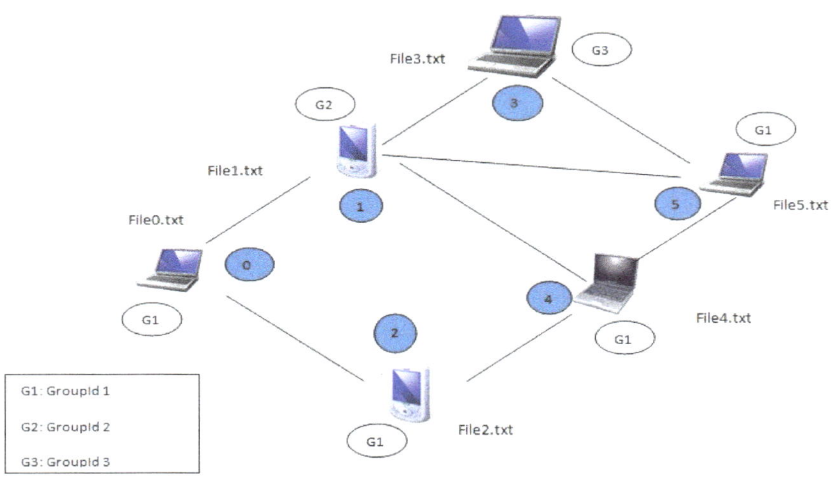

Figure 5: Example illustrating the Algorithm chart

In Figure 5 we show an example of six nodes, assuming that each node in the system has a group-id number, means we are classifying the nodes in our work into different groups, which in our case three groups: G1, G2, and G3. The first group has node 0, node 2, node 4 and node 5. Whereas node 1, node 3 are in G2 and G3 respectively. In our work we make node 0 broadcast a message to all nodes in the group-id specified in the policy file at file0.txt in node node 0, and we call this group-id in this situation a permitted group as shown in the algorithm chart in Figure 6. If file0.txt as in the example has been marked with the group-id 1 that means only nodes in the group G1 can receive the packet. Node 0 will start searching for the adjacent nodes in transmission range. In this example node 0 will find node 1 and node 2 and select them as potential destinations (dest1=n1, dest2=n2). Now node 0 will check if dest1 in the permitted group or not, and do the same to dest2 also. In the algorithm (Figure 6) this depicted as Getgroupid (dest) process and checks if the group-id is permitted. In this example it will be permitted for node 2 as it is in the G1. Node 0 will send to node 2 not only the packet it also sends its policy contained in file0.txt, whereas node 2 will create a packet handler to receive the packet. Once it received the policy, node 2 will be updated according to received policy updating its old policy.

Now, when node 2 at another time wants to broadcast the message again will start and do the same process like node 0. This time, however, node 2 will send to node 4 but not to node 0 because node 0 is the originator of the packet as shown in the algorithm chart in Figure 6, and the system will continue through the same steps for other nodes.

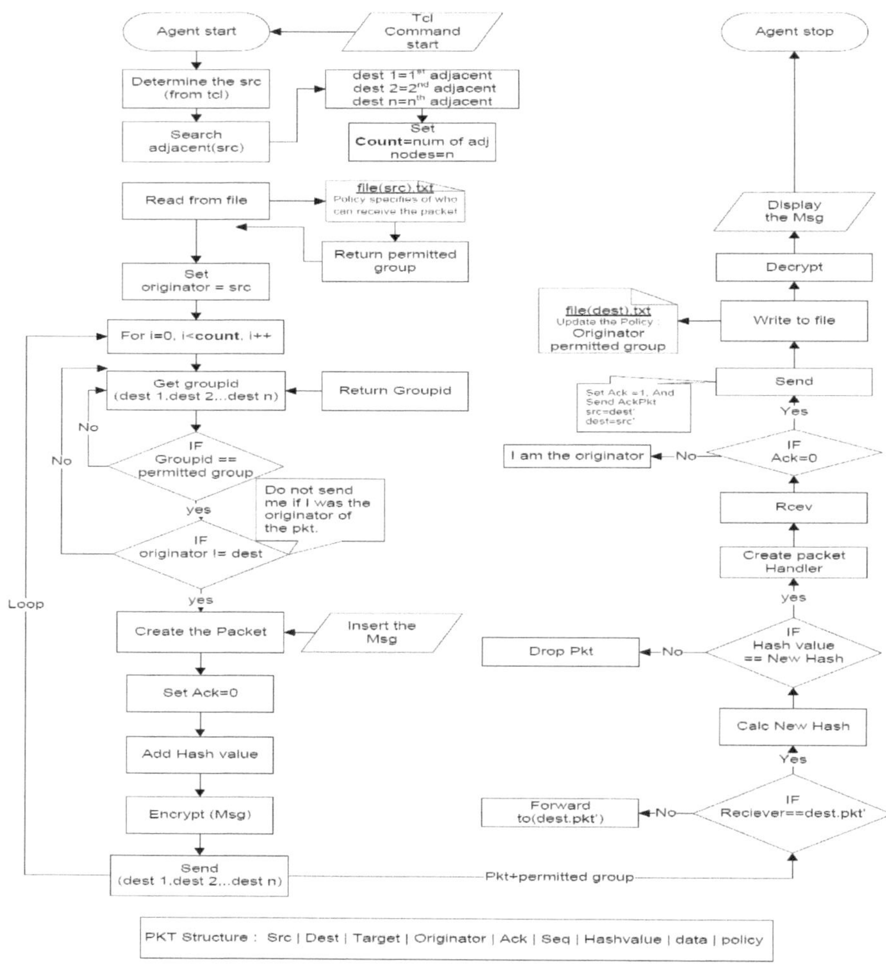

Figure 6: Our Chart Algorithm and packet structure

6. Result and Discussion

In this work we used the Network Simulator (NS-2) which is a real network environment simulator, which showed only intended nodes can receive the packet which has been sent by the source. We simulated our approach into multi variable number of nodes where the originator node disseminates the packet (Data+Policy) to the other nodes. Our result from the tracing file and the nam showed that only nodes in the permitted group-id can receive the packet because of the restriction which has issued from the originator node 'not to send the packet to nodes in different group-id.

We assumed that all nodes in the system are trusted to enforce the policy attached within the packet, and the encryption, digital signature, and the keys management have already been done securely. So in this work we address the stage after the processes mentioned above. So implicitly that means that if a node is trusted to receive a certain information (in the clear) our framework assumes that it also will be trusted to protect this information. In the domains discussed above this appears to be a reasonable assumption that however does not protect against malicious nodes in the network that have infiltrated the system and (wrongly) gained the trust of message originators. Similarly the system does not prevent out-of-band communications and assumes the data is communicated using the provided infrastructure. The protection here is that automated services that provide e.g. situational awareness are trust-enabled to limit the dissemination of information.

We simulated our policy-based agent with a variable number of UDP agents simultaneously to check what happen if all agents are started in the simulation and how the time necessary a packet to be transmitted across a network from source to destination will be affected. In Figure 7 we measured the delay time versus number of cbr traffics which are depicted on the y-axis and x-axis respectively. The result of this figure showed that as the number of cbr traffic increases, the delay time of both agents will increase. We started with 1 cbr traffic, 2, 3 and 4 with and without our policy-based agent to be started at different sources and destinations to measure the average of the delay time between them.

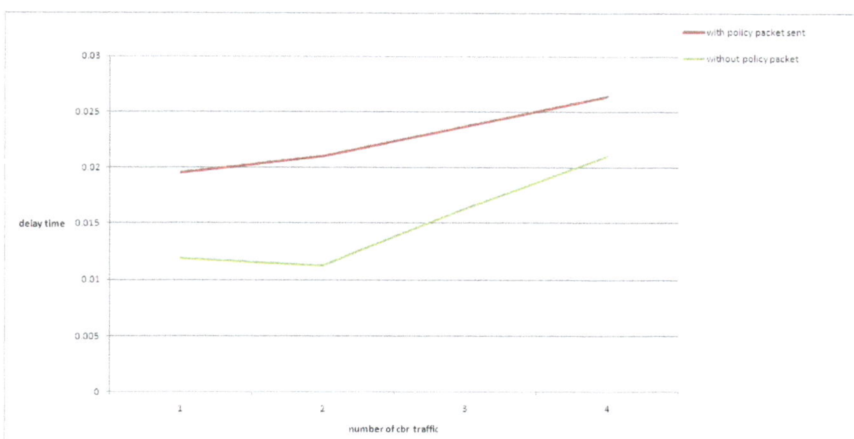

Figure 7: Delay time

It is clear from Figure 7. that there is a delay introduced by sending additional information that contains the policies, but not to a degree that would yield the system unusable.

7. Conclusion

In this paper we conclude that our framework achieves the source policy to send the packet for intended nodes only in the network, on top of that we highlighted the special considerations for security in MANETs and provided an extensive overview of related work and the state of the art in this area. To our knowledge, none of the related work addressed the issue of controlling the information flow in MANETs. We presented a scenario drawn from the military domain, where the impact of this form of confidentiality breach is evident and a real risk. We provided an architecture that addresses this problem by automatically attaching policies to the messages that identify how the information can be used by the receiver, thus limiting the relay of messages based on the originator's confidentiality requirements. We currently assume that all nodes in our system are trusted to correctly enforce the policies that are attached to the message and provide a communication system that includes a policy processing layer dealing with the merging of existing and received policies. However, the assumption that all are trusted is strong. In future work we shall relax this assumption by providing traceability, *viz.* watermarking messages in such a way that they are identified as compromised, they will then lead to a dynamic adjustment of the originator policy based on the resulting trust.

References

[1] Mehul, Ekata, and Vikram Limaye. Security in Mobile Ad Hoc Networks. Handbook of Research in Mobile Business,2nd Ed. Bhuvan Unhelkar. Hershey: IGI Global, 2009. 541-58.

[2] Subir Kumar Sarkar, T. G. Basavaraju, and C. Puttamadappa. *Ad Hoc* Mobile Wireless Networks: Principles, Protocols and Applications. Auerbach Publications, Boston, MA, USA, 2007.

[3] M. Carvalho. Security in mobile *ad hoc* networks. Security Privacy, IEEE, 6 (2): 72 –75, 2008.

[4] W. Li and A. Joshi. Security Issues in Mobile Ad Hoc Networks-A Survey. Department of Computer Science and Electrical Engineering, University of Maryland, 2008.

[5] P. Papadimitratos and Z.J. Haas. Securing mobile *ad hoc* networks. Handbook of *Ad Hoc* Wireless Networks, 2002.

[6] Amitabh Mishra and Ketan M. Nadkarni. Security in wireless *ad hoc* networks. pages 499–549, 2003.

[7] H. Yang, H. Luo, F. Ye, S. Lu, and L. Zhang. Security in mobile *ad hoc* networks: challenges and solutions. Wireless Communications, IEEE, 11 (1): 38–47, 2004.

[8] D. Djenouri, L. Khelladi, and N. Badache. A survey of security issues in mobile *ad hoc* networks. IEEE communications surveys, 7 (4), 2005.

[9] J.L. Burbank, P.F. Chimento, B.K. Haberman, and W.T. Kasch. Key challenges of military tactical networking and the elusive promise of manet technology. Communications Magazine, IEEE, 44(11):39–45, 2006.

[10] L. Zhou and Z.J. Haas. Securing *ad hoc* networks. Network, IEEE, 13 (6): 24–30, 1999.

[11] J.P. Hubaux, L. Buttyán, and S. Capkun. The quest for security in mobile *ad hoc* networks. In Proceedings of the 2nd ACM international symposium on Mobile *ad hoc* networking & computing, pages 146–155. ACM, 2001.

[12] S. Capkun, L. Buttyán, and J.P. Hubaux. Self-organized public-key management for mobile *ad hoc* networks. Mobile Computing, IEEE Transactions on, 2 (1): 52–64, 2003.

[13] Jameela Al-Jaroodi. Security issues at the network layer in wireless mobile *ad hoc* networks at the network layer. Technical report, Faculty of Computer Science and Engineering ,University of Nebraska-Lincoln, Nebraska, USA, 2002.

[14] C. Siva Ram Murthy and B.S. Manoj. *Ad Hoc* Wireless Networks: Architectures and Protocols. Prentice Hall PTR, Upper Saddle River, NJ, USA, 2004.

[15] C.K. Toh. Maximum battery life routing to support ubiquitous mobile computing in wireless *ad hoc* networks. IEEE communications Magazine, 39 (6): 138–147, 2001.

[16] R. Chadha and L. Kant. *Policy-driven mobile ad hoc network management.* Wiley-IEEE Press, 2007.

[17] Y. Zhang and W. Lee. Security in mobile ad-hoc networks. ad hoc Networks, pages 249–268, Ed. Prasant mohapatra Srikanthv. Krishnamurthy. Springer, 2005 .

[18] William Stallings. Cryptography and Network Security: Principles and Practice. Pearson Education, 4rd edition, 2005.

[19] A.J. Menezes, P.C. Van Oorschot, and S.A. Vanstone. Handbook of applied cryptography. CRC, 1997.

[20] M. Blaze, J. Feigenbaum, and J. Lacy. Decentralized trust management. In Security and Privacy, 1996. Proceedings., 1996 IEEE Symposium on, pages 164–173. IEEE, 2002.

[21] Y.G. Desmedt. Threshold cryptography. European Transactions on Telecommunications, 5(4):449–458, 1994.

[22] Y. Desmedt and Y. Frankel. Threshold cryptosystems. In Advances in Cryptology—CRYPTO'89 Proceedings, pages 307–315. Springer, 1990.

[23] K.E. Sirois and S.T. Kent. Securing the nimrod routing architecture. In sndss, page 74. Published by the IEEE Computer Society, 1997.

[24] S. Li and X. Wang, Enhanced security design for threshold cryptography in *ad hoc* network.

[25] B. Wu, J. Chen, J. Wu, and M. Cardei, "A survey of attacks and countermeasures in mobile *ad hoc* networks," Wireless Network Security, pp. 103–135, 2007.

[26] I. Almomani and H. Zedan, End-to-end security solution for wireless mobile ad hoc network (WMANET), IADIS International Conference Applied Computing, 2007, Spain.

[27] A. Al-Bayatti, H. Zedan, and A. Cau, Security solution for mobile ad hoc network of networks (MANON), ICNS '09. Fifth International Conference on in Networking and Services, pp. 255 – 262, April 2009.

[28] S. Pearson and M. Casassa-Mont, Sticky policies: an approach for privacy management across multiple parties, Computer, no. 99, pp. 1–1, 2011.

[29] K. Paul, R. Choudhuri, and S. Bandyopadhyay, Survivability Analysis of Ah Hoc Network Architecture, Proceedings of the IFIP-TC6/European Commission International Workshop on Mobile and Wireless Communication, pp31 – 46, Springer, 2000.

2

DISCRETE COSINE TRANSFORM-II FOR REDUCTION IN PEAK TO AVERAGE POWER RATIO OF OFDM SIGNALS THROUGH μ-LAW COMPANDING TECHNIQUE

Navneet Kaur and Lavish Kansal

Department of Electronics and Communication Engineering
Lovely Professional University, Phagwara
Punjab, India
nv_neet@yahoo.com
Department of Electronics and Communication Engineering
Lovely Professional University, Phagwara
Punjab, India
lavish.15911@lpu.co.in

Abstract

Orthogonal frequency Division multiplexing (OFDM) is the most familiar word in telecommunication and wireless communication systems as it provides enhanced spectral efficiency than Frequency division multiplexing (FDM).Although it is sustaining an orthogonal relation betweencarriers but high peak to average power ratio (PAPR) is one of the main disadvantages of OFDM system.Various PAPR reduction techniques have been used, including techniques based on companding. Incompanding, μ-Law companding has potential to reduce the PAPR of OFDMsignals. μ-Law Companding technique preserves the dynamic range of samples at low amplitudes.A new method named as precoding which is having less complexity compared to the other power reduction techniques is proposed to reduce PAPR. This paper put forward combination of two existing techniques namely μ-Law Companding Transform and Discrete Cosine Transform-II precoding technique. The simulation results show that, the proposed combinedscheme gives better result for PAPR Reduction and results in no distortion.

Keywords

FDM, FDM, PAPR, FFT, IEEE, DCT, CCDF

1. INTRODUCTION

As the demand for high data rate wireless digital communication is rapidly expanding, several approaches such as increasing modulation order or employing multiple antennas at both transmitter and receiver that are reliable for wireless systems have been studied to enhance the spectral efficiency [1,2]. In today's communication systems Orthogonal Frequency Division Multiplexing (OFDM), widespread modulation technique has been considered to achieve high rate data transmission for wireless environment.OFDM system has high spectral efficiency, ease of implementation using the fast Fourier transform (FFT) and simple equalization techniques and also provides immunity to multipath fading, robustness against inter-symbol interference. Therefore it is considered for digital communication systems and also for future broadband applications.

Multicarrier networks such as Frequency Division Multiplexing (FDM) have been around since the late 1950's [3], however due to their implementation complexity transmitted in parallel. Multicarrier systems in wireless applications are less susceptible to channel induced distortions than single carrier systems at corresponding data rates and inefficient use of the frequency band they were restricted to military applications.

In multicarrier transmission, a single data stream bearing a large number of narrow-band subcarriers is transmitted over a number of lower rate subcarriers.Thesesubcarriers are regularly spaced in frequency, domain to avoid the interference. The subcarriers are totally independent of each other because thefrequency spacing and time synchronization of the subcarriers is chosen in such a waythat the subcarriers are orthogonal. By selecting a special set of orthogonal carrier frequencies, high spectral efficiency isobtained because the spectra of the sub-carriers do not overlap. The orthogonality among subcarriers can be maintained by introducing a cyclic prefix over a dispersive channel. The advantage of this structure over single carrier systems is that the extended symbol time due to lower data rate makes the signal less susceptible to effects of the channel such as multipath propagation which introduces Inter Symbol Interference (ISI). Each sub-channel will therefore experience flat fading reducing the equalization complexity in the receiver.

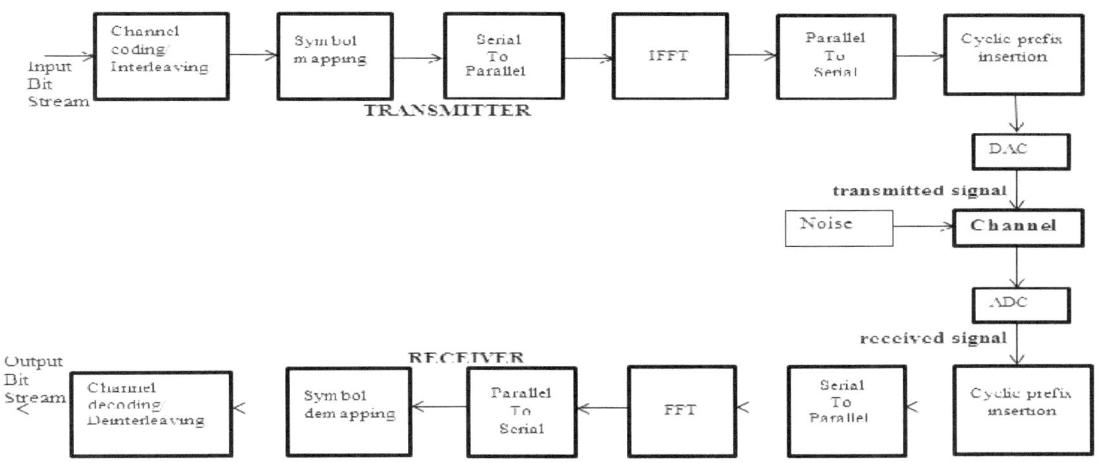

Fig. 1: Block Diagram of OFDM system

In an OFDM scheme shown in fig. 1, the orthogonal and overlapped subcarriers are a large number of orthogonal, overlapping, narrow band sub-carriers are transmitted in parallel.Maintaining total data rates similar to conventional single-carrier modulation schemes in the same bandwidth to modulate each sub-carrier is with a conventional modulation scheme such as Quadrature Amplitude Modulation (QAM) or Phase-Shift Keying (PSK) at a low symbol rate than that required for the whole data stream, but still.In OFDM system a high speed data is splits into a number of parallel lower data rate streams and these low rates data streams are transmitted simultaneously over a number of orthogonal subcarriers.The amount of dispersion in time caused due to multipath delay is reduced as, the symbol duration increases for lower rate parallel sub-carriers. These carriers divide the available transmission bandwidth.

The sub-carriers are separated in such a way so that there is a very compact spectral utilization and each being modulated at a low bit rate. In a conventional frequency division multiplex the carriers are spaced apart to avoid spectral overlap and multipath distortion.

OFDM has progressed to the point where it has now been widely used for various communication applications such as digital audio and video broadcasting, asymmetric digital subscriber line modems, and wireless local area networks systems(WLAN).It has also been adopted as the physical layer modulation scheme for wireless networking standards such as Hiperlan2 in Europe and the Institute of Electrical and Electronic Engineers (IEEE) 802.11a, g standards in the United States.
OFDM systems have a large dynamic signal range with a very high peak to average power ratio (PAPR) which is a one of major drawbacks of the transmitted OFDM signal. As a result, the OFDM signal will be clipped when passed through a nonlinear high power amplifier (HPA) at the transmitter as it only operate in its linear region with sufficient back-off. Thus, HPA with a large dynamic range is required for OFDM systems. These amplifiers are very expensive and are major cost component of the OFDM systems. Thus reduction in the value of the PAPR not only reduces the cost of OFDM systems and reduces the complexity of analog-to-digital (A/D) and digital-to-analog (D/A) converters, but also increases the transmit power, thus, improving received signal-noise-ratio (SNR).

To reduce the PAPR, several techniques have been proposed, which can be divided into three groups.The first group of reduction techniques includes signal distortion techniques, which reduce the peak amplitudes simply by nonlinearly distorting the OFDM signal at or around the peaks. Examples of distortion techniques are clipping [5], peak windowing [6], and peak cancellation. Second, there are coding techniques that use a special code set that excludes OFDM symbols with a large PAPR. The special codes includes Golay complementary sequence [7], Reed Muller code [8], and Hadamard code in this approach [9].The third technique scrambles each OFDM symbol with different scrambling sequences and selecting the sequence that gives the smallest PAPR such as selected mapping (SLM) [10] and partial transmit sequence (PTS) [11].

The organization of this paper is as follow. Section 2 presents formulates the problem of PAPR in OFDM system.The reduction technique based on μ-Law companding is explained in section 3.Section 4 describes the precoding based Discrete Cosine Transform-II technique. PAPR schemes based on combined DCT-II and μ-Law companding technique are proposed in section 5. Computer simulations results are presented in Section 6. Section 6 draws conclusions regarding the existing and proposed combined technique.

2. PAPR IN OFDM SIGNAL

Presence of large number of independently modulated sub-carriers in an OFDM system increasesthe peak value of the system as compared to the average of the whole system. The ratio of the peak to average power value is termed as Peak to Average Power Ratio.At some instant in time coherent addition of independently modulated *N* sub-carriers same phase produces a peak:

$$\text{PAPR}[x(t)] = \frac{P_{PEAK}}{P_{AVERAGE}}$$

$$=10log_{10}\frac{max[|X(n)|^2]}{E[|x_n|^2]} \tag{1}$$

WhereP_{PEAK} represents peak output power,$P_{AVERAGE}$means average output power. E[·] denotesthe expected value, x_n represents the transmitted OFDM signals which are obtained by takingIFFT operation on modulated input symbols is expressed as:

$$x_n=\frac{1}{\sqrt{N}}\sum_{i=0}^{N-1} X_l\, e^{j\frac{2\pi}{N}ni} \tag{2}$$

Where$N$ is the total number of subcarriers and the OFDM symbol duration is T seconds.

An OFDM symbol consists of N subcarriers by the frequency spacing of Δf. The total bandwidth$B$of a symbol will be divided into N equally spaced subcarriers and thus, all the subcarriers are orthogonal to each other within a time interval of length $T=1/\Delta f$. A complex modulation symbol $X_{m,n}$, isused to modulate each sub-carrier independently where m is a time index and n is a subcarrier index.Then within the time interval T the following signal of the *m-th* OFDM block period can be described as:

$$x_m(t)=\frac{1}{\sqrt{N}}\sum_{N=0}^{N-1} X_{m,n}\, b_n(t-mT) \tag{3}$$

Where, b_n(t) is defined as:

$$b_n(\mathrm{t})=\begin{cases}\exp(j2\pi n\Delta f t), & 0\le t\le T\\ 0, & else\end{cases} \tag{4}$$

Where, b_n(t) is a rectangular pulse applied to each subcarrier. The total continuous time signal *x (t)* consisting of all the OFDM block is given as:

$$x(t)=\frac{1}{\sqrt{N}}\sum_{m=0}^{\infty}\sum_{n=0}^{N-1} X_{m,n}\, g_{n(t-mT)} \tag{5}$$

When compared with traditional single carrier systems, the instantaneous output of an OFDM system often has large fluctuations. To handle these large fluctuations the system devices, such as power amplifiers, A/Dconverters and D/A converters, must have large linear dynamic ranges. If the peak signal goes into the non-linearregion of devices at the transmitter,a series of undesirable interference is encountered, such as high out of band radiation and inter-modulationdistortion. Therefore PAPR reduction techniques are requiredfor OFDM systems. The power output has large fluctuations due to which, the high power amplifier (HPA) should have largedynamic range. This results in poor power efficiency.

OFDM systems combine a number of independent sub carriers which causes increase in PAPR. The increase in PAPR is related to the number of sub carriers and their order of modulation. A curve called as Complementary Cumulative Distribution Function (CCDF)is used to present the vital information regarding the OFDM signal to be transmitted. These curves also provide the PAPR data needed by component designer. The main use of power CCDF curves is to identify the power characteristic of the signals which are amplified, mixed and decoded [12]. The plot of relative power levels of signal against their probability of occurrence is called CCDF curve.

This curve illustrates the amount of time the signal remains at or above a given threshold level [13, 14]. The ratio between power level and the average power is expressed in dBas:

$$\text{CCDF}=\text{Prob}(\text{PAPR} > PAPR_o) \tag{6}$$

where $PAPR_o$is the threshold level.

3. μ-Law Companding Technique

In an OFDM system,large signals occur very infrequently therefore itis similar to thespeech signal, so the same companding technique can be used to improve the OFDM transmission performance. One of the companding technique named asμ-Law Companding can effectively reduce the Peak to Average Power Ratio (PAPR) of the transmitted or the companded OFDM signals by transforming the statistics of the amplitudes of these signals into uniform distribution. The key idea of the μ-law companderis that it covers all the encountered amplitudes of the OFDM signal. In μ-law companding, the uniform distribution of the signals can be obtained by compressing the peak signals and expanding the small signals. In the process of companding, the amplitudes of the small signalsget enlarged, while the peaks remain unchanged. Therefore, the average power is increased and thus the PeaktoAverage Power Ratio (PAPR) can be reduced. This technique overcomes the deficiencies that are caused by clipping the original OFDM signal. The other advantage of the companding transform is that, it eliminates the Out of Band Interference (OBI) and can also maintain a constant average power level and also.

The original OFDM signal is converted into the companded signal by using the Mu Companding Transform. The companded signal obtained by using the MuCompanding Transform is given by the equation as:

$$H(x)=\frac{sgn(x)ln(1+\mu|x|)}{\ln(1+\mu)} \tag{7}$$

Where, $H(x)$=Companded Signal obtained byμ-LawCompandingTechnique,
$sgn(x)$=sign function,
x=original OFDM signal,
μ=255 and it is the Mu-Law parameter of the compander, which controls the amount ofcompression.

4. Discrete Cosine Transform-II

The discrete cosine transforms (DCT) is a Fourier-related transform similar to the discrete Fourier transform (DFT). It is a member of a family of sinusoidal unitary transforms. This transform is real, orthogonal and separable with fast algorithms for its computation. They have abetter job of concentrating energy and compression of audio and data. The Discrete Cosine Transform (DCT) was first proposed by Ahmed *et al.* (1974) [15]. It is a Fourier-like transform, which equivalent to a DFT of roughly twice the length, operating on real data with even symmetry. The discrete cosine transform-II is a linear, invertible function $F:R_n \rightarrow R_n$; where$R$ denotes the set of real numbers, or equivalently an $n \times n$ square matrix. The$n$ real numbers$x_{0,\dots,}x_{n-1}$ are transformed into the n real numbers$f_{0,\dots,}f_{n-1}$according to one of the formulas:

$$f_j = \sum_{k=0}^{n-1} x_k \cos\left[\frac{\pi}{n} j \left(k + \frac{1}{2}\right)\right] \quad (8)$$

The DCT-II implies the boundary conditions: x_k is even around k=-1/2 and even around k=n-1/2; f_j is even around j=0 and odd around j=n.

The DCT II is important because it has better computational efficiency, superiority in bandwidth compression of a wide range of signals, its powerful performance in the bit-rate reduction and robust against inter carrier interference (ICI). The precoding based techniques are simple and linear techniques with less complexity to implement without the need of any side information.

5. PROPOSED SCHEME

A new technique named as precoding technique is also used in OFDM systems to reduce the PAPR. Precoding techniques obtain a signal with lower PAPR than in the case of a without precoded OFDM system and it also reduces the interference produced by multiple users. The μ-LawCompanding Transform also causes less spectrum side-lobes as compared to other reduction techniques.In proposed scheme μ-Law companding was combined with precoding technique called Discrete Cosine Transform-II. The combined techniques based OFDM system was shown in fig. 2. In the transmitter, firstly DCT-II is used to transform the parallel data stream, and then the transformed data is processed by the companding unit. If data block passed by DCT-II, before IFFT, the PAPR is reduced, then the PAPR of OFDM signal could be further reduced by companding.

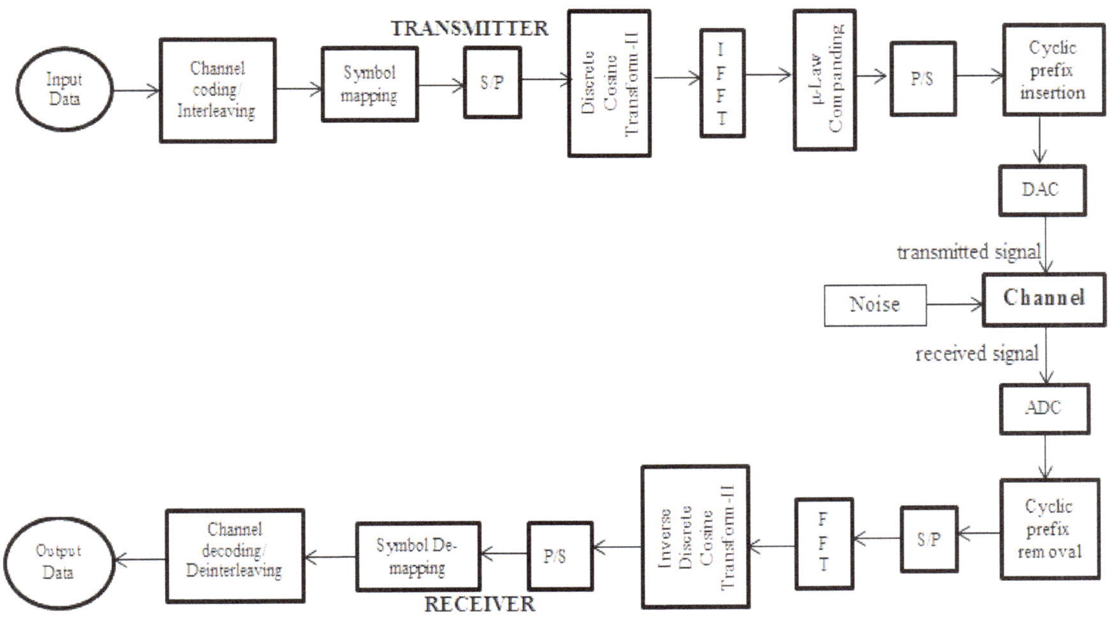

Fig. 2 Block scheme of DCT-II Precoding technique withμ-law companding in OFDM system

After applying the DCT-II on OFDM signal the signal get compressed decreasing the PAPR value. Mucompanding is performed on the I and Q outputs of the IFFT after precoding. As the word length at the IFFT output is decreased, the power consumption and complexity of the DAC/ADC decreases. Also companding at the IFFT output increases the resolution giving a better average signal.

6. SIMULATION RESULTS

To reduce the PAPR, the above discussed has been designed and results are shown in the form of CCDF vs. PAPR plot for different modulations and different values of M. In simulation, an OFDM system is considered under an additive white Gaussian noise (AWGN) channel, where data is randomly generated with subcarriers *N=2400*, then the signal is modulated by *M-PSK* and *M-QAM* respectively. The PAPR reduction capability is measured by the complementary cumulative distribution *(CCDF = Prob (PAPR>PAPR0)*, which indicates the probability that PAPR is above a certain threshold. We compared the simulation results of proposed system with DCT-II precoded OFDM signal and precodedμ-Law companded signal. The results were observed for precoded signal and for precodedμ-Law companded signal.

6.1 M-PSK Modulation

In this section μ-Law companding technique is applied over DCT-II precoded OFDM signal with phase shift keying modulation. We assume CCDF clip rate of $10^{-0.6}$ using M-ary PSK technique for N=2400 and M=16, 32, 64, 128,256, 512 and 1024.

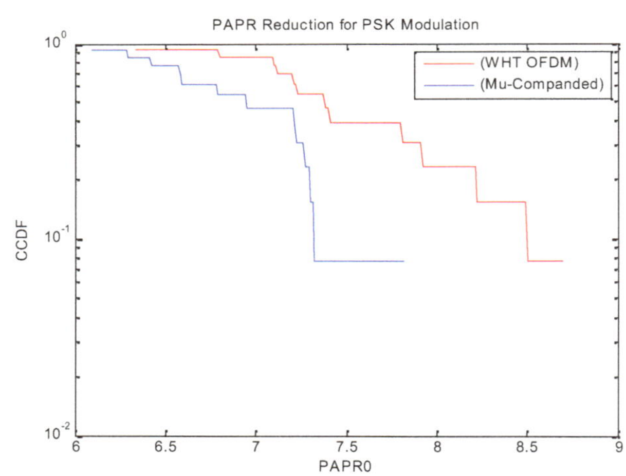

Fig. 3: CCDF of μ-Law companding with proposed DCT-II technique for 16 PSK

For OFDM system CCDF vs. PAPR plots using M=16 over AWGN channel is shown in Fig. 3. The graph gives the clear idea that at CCDF rate of $10^{-0.6}$ the PAPR value decreases by 0.53 dB using PSK modulation. The proposed precoded μ-Law companded technique provides much reduction in PAPR value as compared to DCT-II precoded signal.

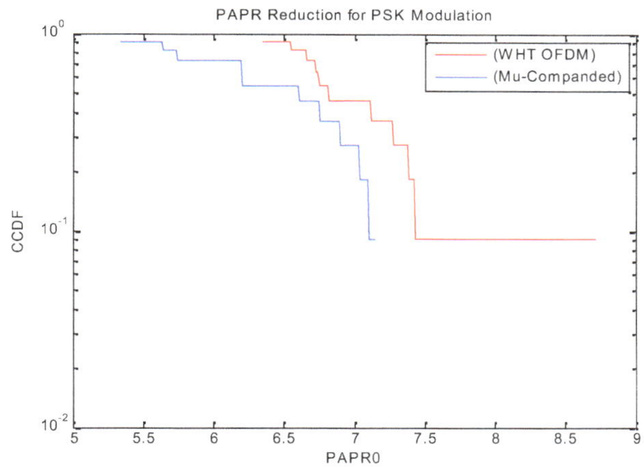

Fig. 4: CCDF of μ-Law companding with proposed DCT-II technique for 32 PSK

For OFDM system CCDF vs. PAPR plots using M=32 over AWGN channel is shown in Fig. 4. The graph gives the clear idea that at CCDF rate of $10^{-0.6}$ the PAPR value decreases by 0.51 dB using PSK modulation. The proposed precodedμ-Law companded technique provides much reduction in PAPR value as compared to DCT-II precoded signal.

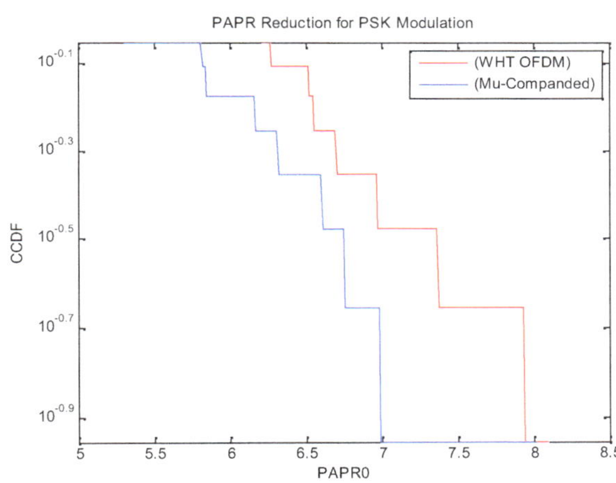

Fig. 5: CCDF of μ-Law companding with proposed DCT-II technique for 64 PSK

For OFDM system CCDF vs. PAPR plots using M=64 over AWGN channel is shown in Fig. 5. The graph gives the clear idea that at CCDF rate of $10^{-0.6}$ the PAPR value decreases by 0.7 dB using PSK modulation. The proposed precoded μ-Law companded technique provides much reduction in PAPR value as compared to DCT-II precoded signal.

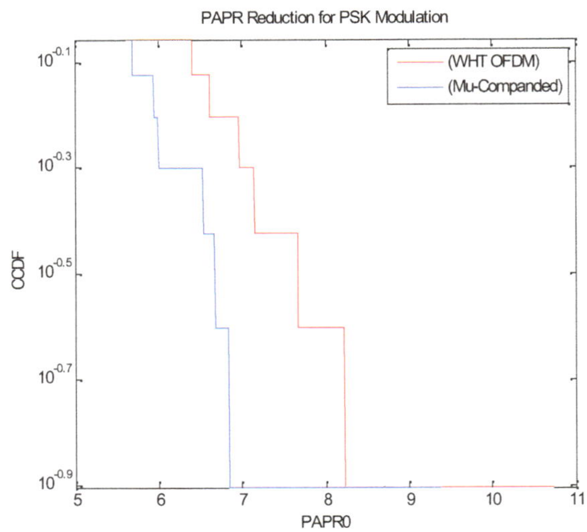

Fig. 6: CCDF of μ-Law companding with proposed DCT-II technique for 128 PSK

For OFDM system CCDF vs. PAPR plots using M=128 over AWGN channel is shown in Fig. 6. The graph gives the clear idea that at CCDF rate of $10^{-0.6}$ the PAPR value decreases by 1.03 dB using PSK modulation. The proposed precoded μ-Law companded technique provides much reduction in PAPR value as compared to DCT-II precoded signal.

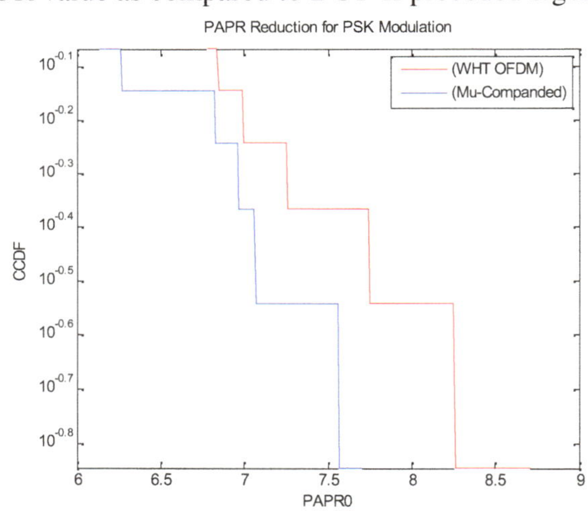

Fig. 7: CCDF of μ-Law companding with proposed DCT-II technique for 256 PSK

For OFDM system CCDF vs. PAPR plots using M=256 over AWGN channel is shown in Fig. 7. The graph gives the clear idea that at CCDF rate of $10^{-0.6}$ the PAPR value decreases by 0.65 dB using PSK modulation. The proposed precoded μ-Law companded technique provides much reduction in PAPR value as compared to DCT-II precoded signal.

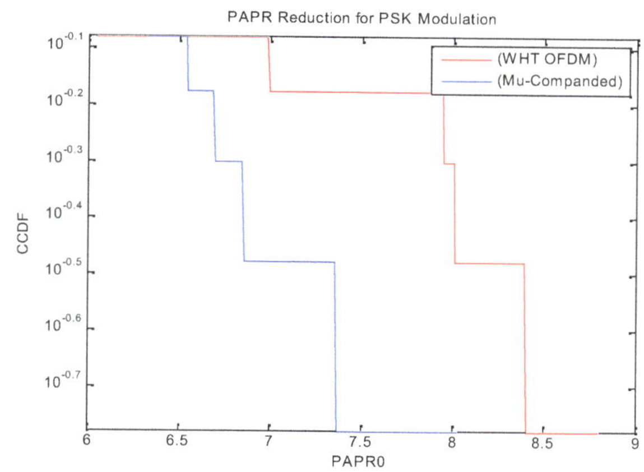

Fig. 8: CCDF of μ-Law companding with proposed DCT-II technique for 512 PSK

For OFDM system CCDF vs. PAPR plots using M=512 over AWGN channel is shown in Fig. 8. The graph gives the clear idea that at CCDF rate of $10^{-0.6}$ the PAPR value decreases by 1.02 dB using PSK modulation. The proposed precoded μ-Law companded technique provides much reduction in PAPR value as compared to DCT-II precoded signal.

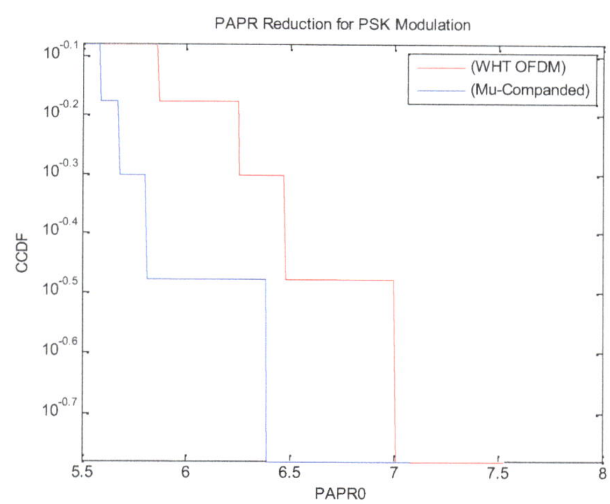

Fig. 9: CCDF of μ-Law companding with proposed DCT-II technique for 1024 PSK

For OFDM system CCDF vs. PAPR plots using M=1024 over AWGN channel is shown in Fig. 9. The graph gives the clear idea that at CCDF rate of $10^{-0.6}$ the PAPR value decreases by 0.62 dB using PSK modulation. The proposed precoded μ-Law companded technique provides much reduction in PAPR value as compared to DCT-II precoded signal.

6.2 M-QAM Modulation

For M=16, 64, 256 and 1024 M-ary quadrature amplitude modulation (QAM) technique is used here at CCDF rate of $10^{-0.6}$.

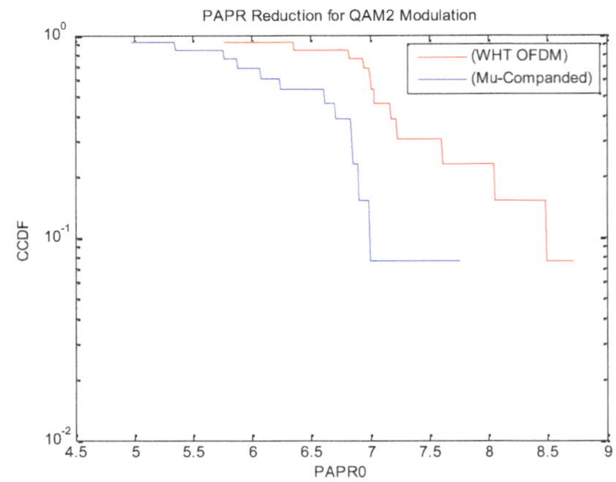

Fig. 10: CCDF ofμ-Law companding with proposed DCT-II technique for 16 QAM

For OFDM system CCDF vs. PAPR plots using M=16 over AWGN channel is shown in Fig. 10. The graph gives the clear idea that at CCDF rate of $10^{-0.6}$ the PAPR value decreases by 0.87 dB using QAM modulation. The proposed precoded μ-Law companded technique provides much reduction in PAPR value as compared to DCT-II precoded signal.

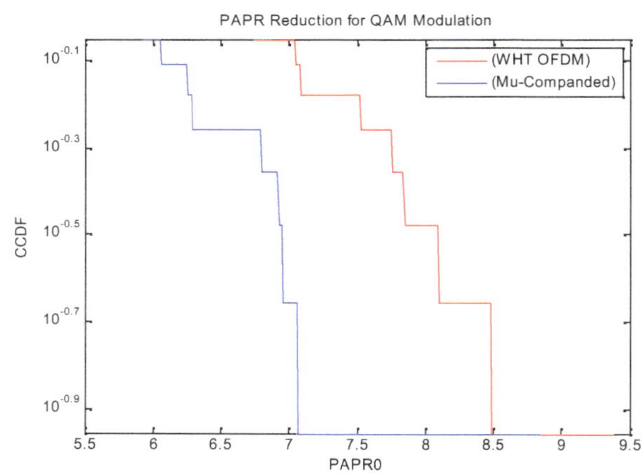

Fig. 11: CCDF of μ-Law companding with proposed DCT-II technique for 64 QAM

For OFDM system CCDF vs. PAPR plots using M=64 over AWGN channel is shown in Fig. 11. The graph gives the clear idea that at CCDF rate of $10^{-0.6}$ the PAPR value decreases by 0.83 dB using QAM modulation. The proposed μ-Law companded technique provides much reduction in PAPR value as compared to DCT-II precoded signal.

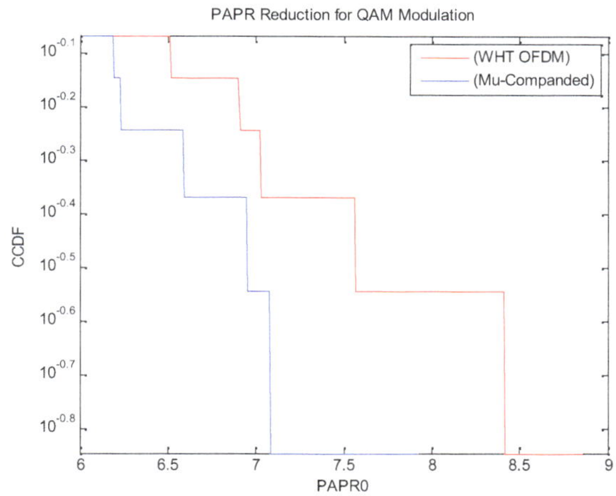

Fig. 12: CCDF of µ-Law companding with proposed DCT-II technique for 256 QAM

For OFDM system CCDF vs. PAPR plots using M=256 over AWGN channel is shown in Fig. 12. The graph gives the clear idea that at CCDF rate of $10^{-0.6}$ the PAPR value decreases by 1.23 dB using QAM modulation. The proposed precoded µ-Law companded technique provides much reduction in PAPR value as compared to DCT-II precoded signal.

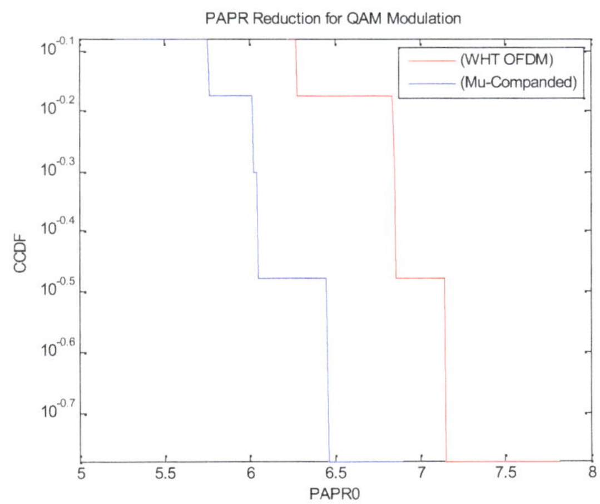

Fig. 13: CCDF of µ-Law companding with proposed DCT-II technique for 1024 QAM

For OFDM system CCDF vs. PAPR plots using M=1024 over AWGN channel is shown in Fig. 13. The graph gives the clear idea that at CCDF rate of $10^{-0.6}$ the PAPR value decreases by 1.3 dB using QAM modulation. The proposed precoded µ-Law companded technique provides much reduction in PAPR value as compared to DCT-II precoded signal.

7. CONCLUSIONS

In this paper, a new idea about the performance of the µ-Law companding technique combined with precoded technique is presented for the reduction of PAPR value. The PAPR value can be reduced for different modulation formats and for different values of M by using proposed scheme without increasing the system complexity. The scheme shows superiority in bandwidth compression of signals and powerful performance in the bit-rate reduction. This system produce the lowest PAPR and is efficient, signal independent, distortion less and does not causes any spectrum side-lobes. The simulation results shows, that the proposed precoded combinedμ-law companding method displays a better PAPR reduction performance than the simple DCT-II precoded OFDM signal. Thus, it is concluded that the proposed scheme reduces PAPR effectively compared with precoded DCT-II transform.

References

[1] H. Jiang & P. A. Wilford, "A hierarchical modulation for up grading digital broadcasting systems," IEEE Transaction on Broadcasting, Vol. 51, pp. 222-229, 2005.

[2] P. W. Wolniansky, G. J. Foschini, G. D. Golden & R. A. Valenzuela, "V-BLAST: an architecture for realizing very high data rates over the rich- scattering wireless channel," In Proceeding of International symposium on Signals, Systems and Electronics, pp. 259-300, 1998.

[3] J. A. C. Bingham, "Multicarrier modulation for data transmission: An idea whose time has come", IEEE Communication Magazine, Vol. 28, Issue: 1, pp. 5-14, 1990.

[4] R. Gross & D. Veeneman, "Clipping distortion DMT ADSL systems," IEEE Electronics Letters, Vol. 29, pp. 2080- 2081,1993

[5] T. Jiang & Y. Wu, "An Overview: Peak to Average Power Ratio Reduction Techniques for OFDM Signals," IEEE Transactions on Broadcasting, Vol. 54, Issue: 2, pp. 257-268, 2008

[6] F. Fiedler, J. Jedwab& M. G. Parker," A framework for the construction of Golay sequences," IEEE Information Theory, Vol. 54, pp. 3114-3129, 2008.

[7] J. A. Davis & J. Jedwab, "Peak to mean power control and error correction for OFDM transmission using Golay sequences and Reed-Muller codes," Electronics Letters, Vol. 33, pp. 267-268, 1997.

[8] M. Park, J. Heeyong, J. Cho, D. Hong & C. Kang,"PAPR Reduction in OFDM Transmission Using Hadamard Transform," IEEE International Conference on Communications, Vol. 1, pp. 430-433, 2000.

[9] R. W. Bauml, R. F. Fischer & J. B. Huber, "Reducing the Peak to Average Power Ratio of Multicarrier Modulation by Selected Mapping," Electronics Letters, Vol. 32, Issue: 22, pp. 2050-2057, 1996.

[10] H. Breiling, S.H. H. Muller Weinfurtner& J.B. Huber, "SLM peak power reduction without explicit side information," IEEE Communication Letters, Vol. 5, Issue: 6, pp. 239-241, 2001.

[11] S. H. Muller and J. B. Huber, "OFDM Ratio by Optimum Combination of Partial Transmit Sequences," IEEE Electronic Letters, Vol. 33, Issue: 5, pp. 368-369, 1997.

[12] Agilent Technologies Application Note, "Characterizing Digitally ModulatedSignal with CCDF Curves", 2000.

[13] R.Van Nee and A. de Wild, "Reducing the peaktoaverage power ratio of OFDM,"IEEE Proceeding on Vehicular Technology Conference, pp. 2072-2076, 1998.

[14] H. Ochiai and H. Imai, "On the distribution of thepeak to average power in OFDM signals," IEEE Transactions on Communications, Vol.49, pp.282-289, 2001.

[15] N. T. Ahmed, Natarajan& K. R. Rao, "Discrete Cosine Transform," IEEE Transactions on Computer, Vol. 23, pp. 90-93,1974.

[16] L. Kansal, A. Kansal& K. Singh, "Analysis of Different High Level Modulation Techniques for OFDM System," International Journal of VLSI and Signal Processing Applications, Vol. 1, Issue: 2, 2011.

[17] L. Kansal, A. Kansal& K. Singh, "Performance Analysis of MIMO-OFDM System Using QOSTBC Code Structure for M-PSK," Signal Processing: An International Journal, Vol. 5, Issue: 2, 2011.

[18] N. Kaur& L. Kansal, "Reducing the Peak to Average Power Ratio of OFDM Signals through Walsh Hadamard Transform," Global Journal of Researches in Engineering, Vol. 13, Issue: 1, 2013.

[19] N. Kaur& L. Kansal, "Peak to Average Power Ratio Reduction of OFDM Signal by Combining Clipping withWalsh Hadamard Transform,"International Journal of Wireless& Mobile Networks,Vol. 5, Issue: 1, 2013.

Adaptive Subcarrier and Bit Allocation for Downlink OFDMA System with Proportional Fairness

Mr. Sudhir B. Lande[1], Dr. J. B. Helonde[2], Dr. Rajesh Pande[3] S.S.Pathak[4]
[1] Asst. Prof. K.I.T.S. Ramtek Dist- Nagpur , India
[1]landeeed@yahoo.co.in
[2]Principal I T M Kamthee Dist- Nagpur , India
,[2]_helondejb@yahoo.com
[3]Prof. & Vice-Principal R. C. O. E. M. Nagpur, India
[3]panderaj@yahoo.com
[4]Prof. in E&ECE dept. IIT Kharagpur. India
[4]sspathak@iit.ernet.in

ABSTRACT

This paper investigates the adaptive subcarrier and bit allocation algorithm for OFDMA systems. To minimize overall transmitted power, we propose a novel adaptive subcarrier and bit allocation algorithm based on channel state information (CSI) and quality state information (QSI). A suboptimal approach that separately performs subcarrier allocation and bit loading is proposed. It is shown that a near optimal solution is obtained by the proposed algorithm which has low complexity compared to that of other conventional algorithm. We will study the problem of finding an optimal sub-carrier and power allocation strategy for downlink communication to multiple users in an OFDMA based wireless system. Assuming knowledge of the instantaneous channel gains for all users, we propose a multiuser OFDMA subcarrier, and bit allocation algorithm to minimize the total transmit power. This is done by assigning each user a set of subcarriers and by determining the number of bits and the transmit power level for each subcarrier. The objective is to minimize the total transmitted power over the entire network to satisfy the application layer and physical layer. We formulate this problem as a constrained optimization problem and present centralized algorithms. The simulation results will show that our approach results in an efficient assignment of subcarriers and transmitter power levels in terms of the energy required for transmitting each bit of information, to address this need, we also present a bit loading algorithm for allocating subcarriers and bits in order to satisfy the rate requirements of the links.

KEYWORDS

Orthogonal frequency division multiplexing, OFDMA, QSI, QoS, Dynamic resource allocation, Water-filling, SNR.

1. INTRODUCTION

Orthogonal Frequency Division Multiple Access (OFDMA) is an attractive multiple access scheme for future wireless and mobile communication systems, which has been developed to support a variety of multimedia applications with different Quality-of-Service (QoS) requirements. OFDMA builds on Orthogonal Frequency Division Multiplexing (OFDM), which is immune to inter-symbol interference and frequency selective fading, as it divides the frequency band into a group of mutually orthogonal subcarriers, each having a much lower bandwidth than the coherence bandwidth of the channel. OFDMA can be readily implemented in OFDM systems by assigning each user a set of subcarriers. This allocation can be performed in a number of ways. The simplest method is a static allocation of subcarriers to each user. An improvement upon static allocation is dynamic subcarrier allocation, based on channel state conditions OFDMA is essentially a hybrid of FDMA and TDMA: Users are dynamically assigned subcarriers (FDMA) in different time slots (TDMA).The advantages of OFDMA start with the advantages of single-user OFDM in terms of robust multipath suppression and frequency diversity. In addition, OFDMA is a flexible multiple-access technique that can accommodate many users with widely varying applications, data rates, and QoS requirements. Because the multiple access is performed in the digital domain, before the IFFT operation, dynamic and efficient bandwidth allocation is possible[1,2,3]. This allows sophisticated time- and frequency- domain scheduling algorithms to be integrated in order to best serve the user population

1.1 OFDMA Overview

OFDMA technology takes an available wideband channel and divides it in the frequency domain into multiple narrower channels, each with its own subcarrier . This frequency division is assumed to be fine enough such that these subchannels experience only flat (narrowband) fading. Users in an OFDMA system are each assigned distinct and non-overlapping subsets of the subcarriers, and communicate with the base station over their dedicated set of subchannels.
The complexity arises in the allocation of subcarriers to users. Firstly, users in an OFDMA system will generally experience different fading environments, and thus the signal-to-noise ratio (SNR) and received power variations seen by one user will differ from that seen by another. Secondly, different users have different QoS requirements, dictating the SNRs and received powers required at their receivers. Thirdly, there may either be total power constraints or requirements to minimize consumed power. Lastly, wireless channels are subject to constant change in real life, so resource allocation techniques must be fast enough to keep up with changing user environments if real-time applications are to be feasible. Now, in any multi-user system, there is a distinction between the uplink and the downlink. The former describes user-to-base-station communications, whereas the latter describes base-station-to-user communications[4]. These two cases are different due to fundamental differences between these channel models. As a result, resource allocation algorithms particularly those for power will be different for these two cases. Power allocation algorithms will not be presented unless inseparable from the subcarrier allocation algorithms, but even then, the paper will focus only on analysis of the subcarrier allocations.

1.2 Multiuser Diversity and Adaptive Modulation

In OFDMA, the subcarrier and the power allocation should be based on the channel conditions in order to maximize the throughput. In this section, we provide necessary background discussion on the key two principles that enable high performance in OFDMA: multiuser diversity and adaptive modulation. Multiuser diversity describes the gains available by selecting a user or sub- set of users having "good" conditions. Adaptive modulation is the means by which good channels can be exploited to achieve higher data rates. In multi-user environment, each user is dynamically assigned to a subset of subcarriers in each frame, to take advantage of the fact that at any time instance, channel responses are different for different users and at different subcarriers. This capability of OFDMA systems enables the network to perform a flexible radio resource management, such as dynamic subcarrier assignment (DSA), adaptive power allocation (APA), and adaptive modulation and coding (AMC) scheme to improve the system performance significantly under different traffic loads and time-varying channel conditions. We consider multiuser systems where multiple users are allowed to transmit simultaneously on different subcarriers per OFDM symbol. Mobile users on certain OFDM sub-channels may experience deep frequency-selective fading in a multipath propagation environment. Since each user may have a different sub-channel impulse response, a poor sub-channel for one user may be a good sub-channel for another user. Clearly, if a user who suffers from poor sub-channel gain can be reassigned to a better sub channel the total throughput can be increased. This is also known as multi-user diversity. Spatial reuse of channels, performed to increase the network capacity and conserve scarce network resources, should consider physical layer characteristics. Also, the medium access control (MAC) scheme needs to be sensitive to the channel quality experienced by different transmitters and multiuser diversity is exploited [5,6]. The goal of our work is to study the joint optimization of the physical and MAC layers. However, we will focus on subcarrier allocation and bit loading for OFDMA. Orthogonal Frequency Division Multiplexing (OFDM) is a spectrally efficient digital modulation technique. The spectrum of interest is divided into a number of parallel orthogonal narrowband sub-channels (known as subcarriers). When different subcarriers experience different channel gains and interference levels, adaptive bit loading is performed to enhance the throughput for a given transmission power limit.

2. PROBLEM FORMULATION

The aim is to allocate subcarrier at the bit loading algorithm using OFDMA for real time traffic. Here our objective is to assign a set of subcarriers to each link and load bits on each assigned subcarrier. The factors to be considered are the frequency-dependent path loss parameters and the interference due to other links that have already been assigned those subcarriers. Our algorithms are general and will be design to work for satisfying application and physical layer by improving CSI and QSI. Subcarrier allocation is performed at the base station and the users are notified of the carriers chosen for them. After the allocation, each user performs power allocation and bit loading across the subcarriers allocated to it to find the transmission power In this paper we will examine algorithms which fall between these three extremes. Intuitively, we are separating the problem into three stages:

- **Subcarrier Allocation:** we are going through an adaptive multiuser subcarrier allocation scheme where the subcarriers are assigned to the users based on instantaneous channel information.
- **Power Allocation:** Using the channel information, the transmitter applies the combined subcarrier, bit, and power allocation algorithm to assign different subcarriers to different users and the number of bits/OFDM symbol to be transmitted on each subcarrier.
- **Bit allocation:** By this optimal allocation of bits the overall transmit power will be minimized.

This paper is organized as follows: The section 3 deals system model of OFDMA systems. Certain existing subcarrier and power allocation algorithm are analyzed and presented in section 4 Section 5 deals with bit allocation algorithms for OFDM system. Section 6 deals with proposed subcarrier & sub-power allocation algorithm for OFDMA systems and section 7 analyses results of the proposed algorithm

3. OFDMA SYSTEM MODEL

A schematic diagram of an OFDMA-based system used in this paper is shown in Fig 1. At the transmitter, the serial data streams from the users are fed into the OFDM transmitter block. Using the channel information from all users, the subcarrier and bit/power allocation algorithm is applied to assign different subcarriers to different users. Here, it is assumed that a subcarrier at a particular time is not being shared among users. The power allocated to each subcarrier is also determined in the process[6].

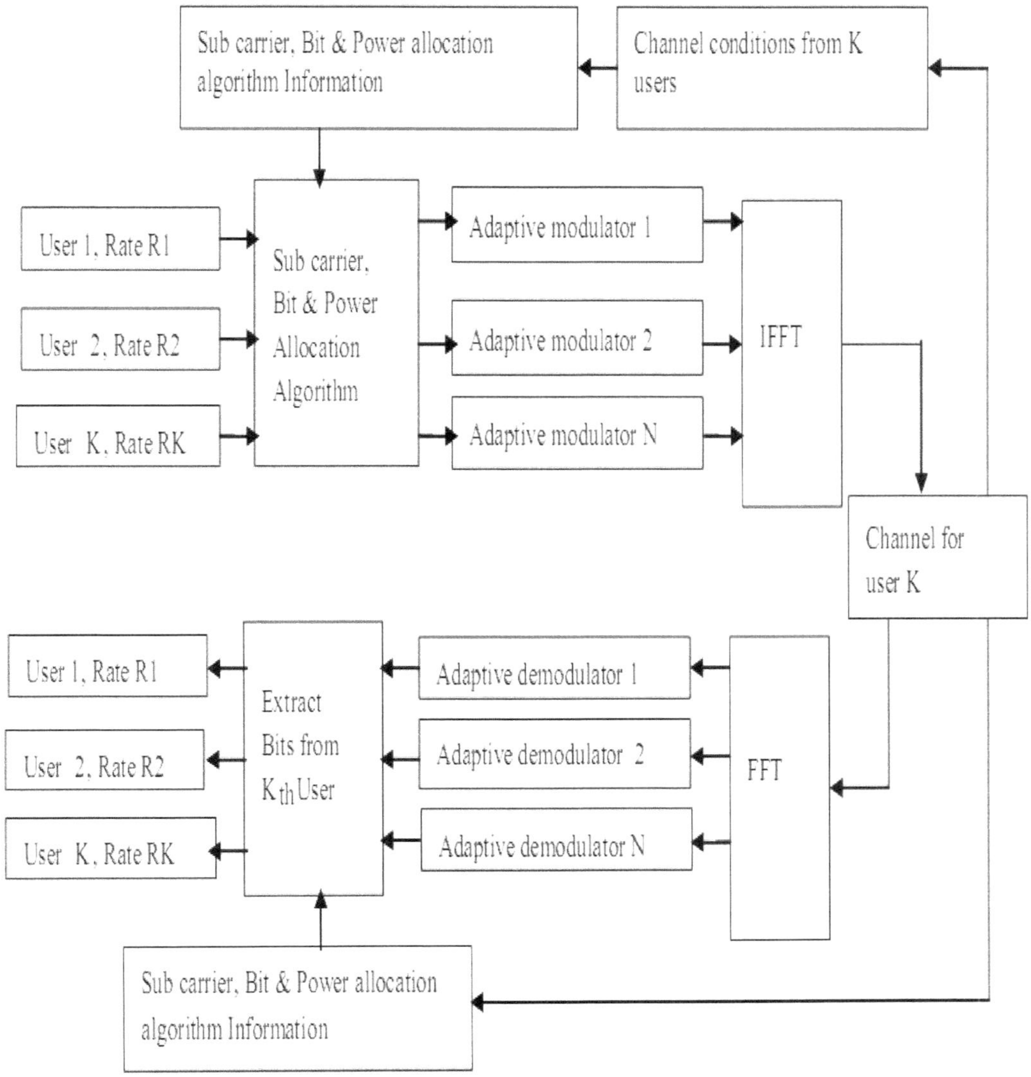

Figure 1: System model for multiuser OFDMA system [6]

This information is used to configure the transmitter. At the receiver, the subcarrier and power allocation information is used to configure the subcarrier selector and decoder to extract the data from the subcarriers assigned to the users. Throughout this paper, certain variables are used to denote the parameters of resource allocation as explained below. It is assumed that a total of K users in the system sharing N sub channels with total transmit power constraint P_{TOT}. The main objective is to optimize the sub channel and power allocation to achieve the highest sum error-free capacity under the total power constraint. It is assumed that each user experiences independent fading and the channel gain of user k in subcarrier n is denoted as $h_{k,n}$, with additive white Gaussian noise (AWGN) $\sigma^2 = N_o \frac{B}{N}$ where N_o is the noise power spectral density. The corresponding sub-channel-to-noise ratio is thus denoted as $H_{k,n} = h_{k,n}^2/\sigma^2$.

4. RESOURCE ALLOCATION ALGORITHMS FOR OFDMA SYSTEMS

To maximize the total capacity, the transmission power needs to be distributed optimally under the total power constraint. The following optimization problem is formulated to determine the optimal power allocation. Mathematically, the optimization problem considered in this paper is formulated as[7]

$$\max \rho_{k,n} \sum_{k=1}^{k} \sum_{n=1}^{N} \frac{\rho_{k,n}}{N} \log_2 \left[1 + \frac{P_{k,n} h_{k,n}^2}{N_o \frac{B}{N}} \right] \qquad (1)$$

Subject to following constraints

$$\sum_{k=1}^{k} \sum_{n=1}^{N} p_{k,n} \leq P_{TOT}$$

$$p_{k,n} \geq 0 \forall k, n$$

$$p_{k,n} = \{0,1\} \forall k, n$$

$$\sum_{k=1}^{k} p_{k,n} - 1 \forall n$$

where, B is the total available bandwidth, $P_{k,n}$ is the power allocated for user k in the subchannel n , $\rho_{k,n}$ is either 1 or 0, indicating whether subchannel n is used by user k or not. The capacity for user k, denoted as R_k , is defined as

$$R_k = \sum_{n=1}^{N} \frac{\rho_{k,n}}{N} \log_2 \left[1 + \frac{p_{k,n} h_{k,n}^2}{N_o \frac{B}{N}} \right] \qquad (2)$$

4.1 Root-Finding Method of Subcarrier and Power Allocation

The root-finding method refers to the adaptive resource allocation proposed in. The name stems from a non-linear equation used to determine the total power assigned to the first user. In root-finding method, the determination of subchannel and power assignment was handled separately. This reduces the computational complexity of solving both variables simultaneously. The proportionality between the users is strictly maintained in this algorithm. A subchannel is initially assigned to the user who has the highest equalized channel gain in that subchannel. After all the subchannels are assigned, the capacity for each user is calculated, assuming equal power distribution. At each iteration, the user with the lowest proportional capacity has the option to pick which subchannel to use, provided the user should have not met their requirements. The subchannel allocation algorithm is suboptimal because equal power distribution in all subchannels is assumed. To assign power over all the subchannels, the root-finding method first determines the total power for each user and then breaks it down to the individual subchannels[8,9]. The total power of every user is calculated by using the equation

$$Pk, total = c_k(P_{1,tot})d_k \qquad (3)$$

The final step in the power allocation is to perform water filling across the subcarriers for each user

$$P_{K,N} = P_{K,1} + \frac{H_{k,n} - H_{k,1}}{H_{k,n} H_{k,1}} \qquad (4)$$

4.2 Linear Method of Subcarrier and Power Allocation

Wong *et al* algorithm is almost similar to Shen *et al* . algorithm with minor difference such as use of unallocated subcarriers and low complexity power allocation algorithm. There are four steps in subchannel allocation algorithm. Step 1 determines the number of subcarriers N_k to be initially assigned to each user. step 2 assigns the subcarriers to each user with highest channel-to-noise ratio. Step 3 picks the user with lowest rate to allocate subchannel at the same time the proportionality between the users are roughly maintained. The step 4 assigns the remaining N * subcarriers to the best users for them, wherein each user can get at most one unassigned subcarrier. This is to prevent the user with the best gains to get the rest of the subcarriers. This policy balances achieving proportional fairness while increasing overall capacity. The linear

$$Pk = \begin{cases} \dfrac{(Ptot - \sum_{k=2}^{k} \frac{bk}{akk})}{(1 - \sum_{k=2}^{k} \frac{1}{akk})} & for\ k \to 1 \\ \dfrac{(bk - P1)}{akk} & for\ k \to 2,3, \dots k \end{cases} \qquad (5)$$

equations involved in the power allocation can be easily solved because of its well-ordered symmetric and sparse structure. To solve the above equation, LU factorization (factorization of a matrix in to lower and upper triangular matrices of the same size respectively) is performed on the coefficient matrix. Finally, using forwards-backwards substitution, the individual user powers are given by The power of individual subchannel is found using the water-filling algorithm[10].

4.3 Joint Subcarrier and Power Allocation Algorithm

The algorithm simplifies the optimization problem into one that has N optimization parameters by assuming equal power allocation to all subcarriers, i.e

$$P_{k,n} = \begin{cases} \dfrac{Ptot}{N} & for \to n \in A \quad for\ all\ n = 1,2, \dots, N \\ 0 \to otherwise & and\ k = 1,2, \dots K \end{cases} \qquad (6)$$

In the joint allocation, optimization of the $N + K$ parameters is carried out by alternating between subcarrier and power allocation. The water filling is used for each user. However, unlike, water filling for each user plays a crucial role in deciding the subcarrier allocation. When a subcarrier is allocated to a user, the power allocated to the user is incremented by P total/ N, i.e., the power allocated to each user is proportional to the number of subcarriers currently allocated to the user. The user's rate is also updated assuming that water filling is used. This updated rate information is used in the allocation of the remaining subcarriers[11,12].

5. BIT ALLOCATION ALGORITHM FOR OFDMA SYSTEMS

The algorithm in is aimed at the maximization of data rate under the constraint of total transmit power and BER. The algorithm is based on the water-filling approach, which is known as optimal to maximize data rate under the constraint of total transmit power. In the given algorithm, however, the water-filling power allocation is not fully performed. Instead, by adjusting only the water-filling level needed in the water-filling power allocation, we may adapt transmit power and number of bits for each subchannel with low computational complexity, and the total number of loaded bits in an OFDM symbol can be maximized while satisfying the constraints. The number of loaded bits, b_m for the mth subchannel may be obtained by

$$b_m = log_2\left(1 + \frac{s_m g_m}{\sigma^2 \Gamma}\right) \qquad (7)$$

$$\bar{b}_m = round(b_m) \qquad (8)$$

where, S_m and g_m denote the transmit power allocated to the m^{th}subchannel, and the power gain of the mth subchannel, respectively. σ is the variance of the additive white Gaussian noise. Γ represents the SNR gap which is a function of the target BER and a channel coding scheme. Note that the number of bits for each subchannel increases with λ, as seen from

$$b_m = [log_2(\lambda_{g_m})]^+ \qquad (9)$$

where $[x]^+ = max\{x, 0\}$. The increase of b_m with λ indicates the increase of $\bar{b}$ and $\hat{s}$ as well from (7). Therefore, to maximize the data rate of the system, the water-filling level λ should be kept as large as possible while satisfying the total transmit power constraint. Consequently, by adjusting λ, transmit power and number of bits for each subchannel can be adapted iteratively, and the number of loaded bits in an OFDM symbol can be maximized while satisfying the total transmit power constraint and BER requirement. At each iteration, the water-filling level λ is adjusted as

$$\lambda = \lambda + \mu \frac{1}{M_{ON}} \cdot \frac{1}{\sigma^2 \Gamma}\left(S - \sum_{m=1}^{M} s_m\right) \qquad (10)$$

where μ is a step size; M_{on} denotes the number of turned-on subchannels which are allowed to transmit data bits. The transmit power for the mth subchannel is calculated as

$$S_m = \frac{\sigma^2 \Gamma}{g_m}\left(2^{b_m} - 1\right) \qquad (11)$$

b_m and $\bar{b}_m$are determined by only the value of the water-filling level, λ, when the channel power gain is available. Therefore, whole water-filling power allocation is not required as in, but only the value of λ is adjusted to satisfy the total transmit power constraint given as

$$\sum_{m=1}^{M} s_m \leq S.$$

6. PROPOSED RESOURCE ALLOCATION ALGORITHM FOR OFDMA SYSTEMS

6.1 Number of sub-carriers determination

- **Inputs**: Each user's bit rate constraint and average channel gain for each user
- **Output**: Number of sub-carriers each user gets assigned
- Two types of sub-carriers: Minimum required sub-carrier and Extra sub-carrier
- Minimum required sub-carriers are to fulfil the user's bit rate constraint in the case that maximum amount of bits will be transmitted in each sub-carrier
- Extra sub-carriers will share bits with minimum required sub-carriers so that the loaded bits in each sub-carrier can be reduced and with an adaptive modulation scheme transmission power to all user can decrease
- No free sub-carrier left

6.2 Sub-carrier allocation

- Inputs: Channel State Information for each user and number of sub-carriers each user gets assigned
- Output: sub-carriers allocation

Phase 1: Constructive initial allocation

1. List the sub-carriers for each user in descend order according to channel gain
2. Check sub-carriers user by user if the number of sub-carrier each user gets is achieved or the sub-carrier has already been assigned to some users
3. If both are NO, assign the sub-carrier to this user, otherwise skip this user to next user

Phase 1 may achieve only a local minimum but not total minimum transmission power

Phase 2: Iterative improvement

- For every iteration, swap a pair of sub-carriers allocated to two users such that the result power can be reduced further
- Power reduction factor is the cost function in order to select the pair of users and pair of sub-carriers which can reduce power most
- Iteration is over when the maximum possible power reduction is less than zero

6.3 Bit loading

- Inputs: Sub-carriers allocation, channel gain and bit rate constraint
- Output: Bits loaded to achieve each user's bit rate constraint

1. Each time selecting the sub-carrier that requires the least additional power to add one more bit
2. Check if the maximum amount of bits loaded in this sub-carrier has already been achieved and if this user's bit rate constraint has been fulfilled
3. If both are NO, loading one more bit to this sub-carrier, otherwise selecting the sub-carrier which requires second least additional power and repeat 2

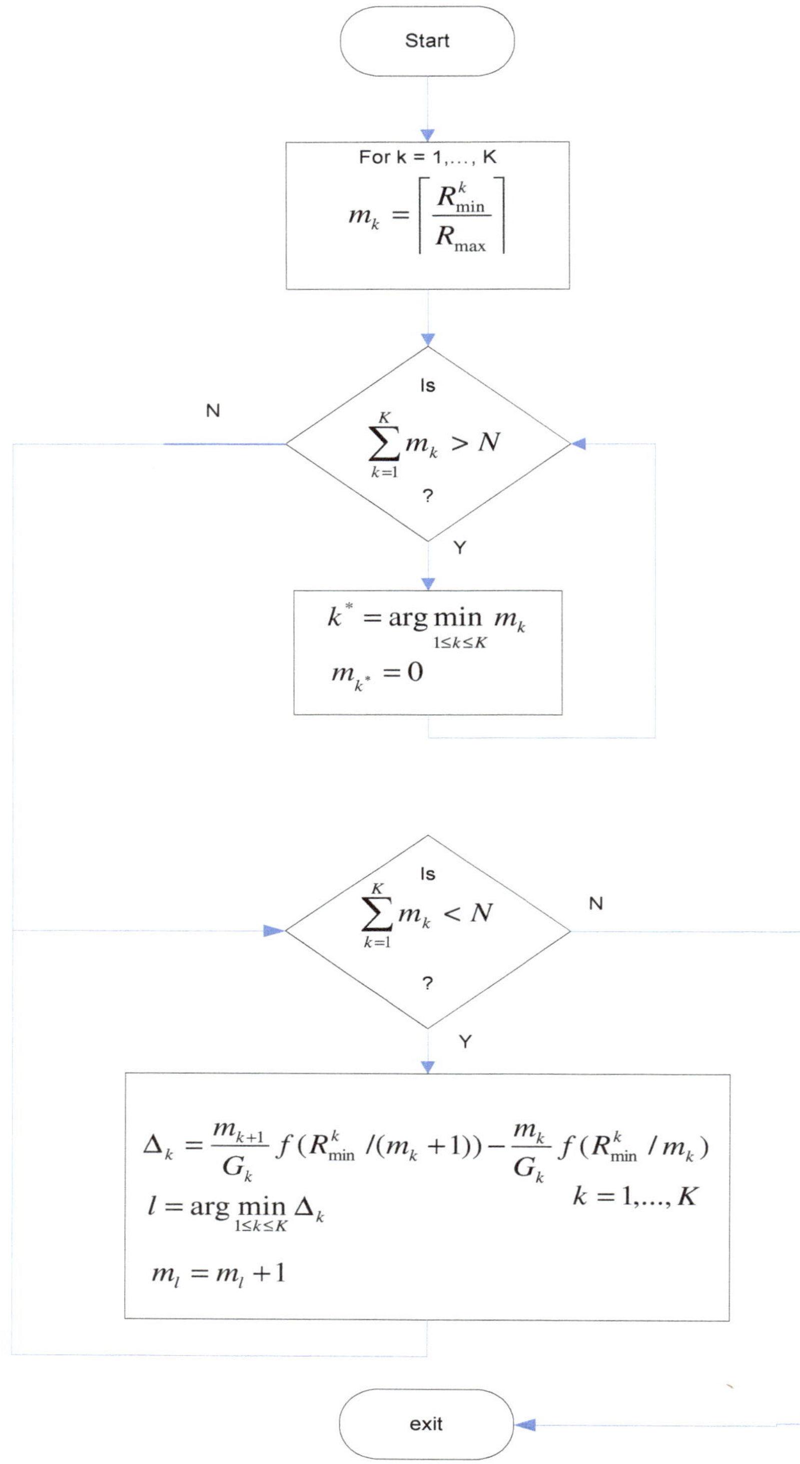

Figure 2 Subcarrier determination

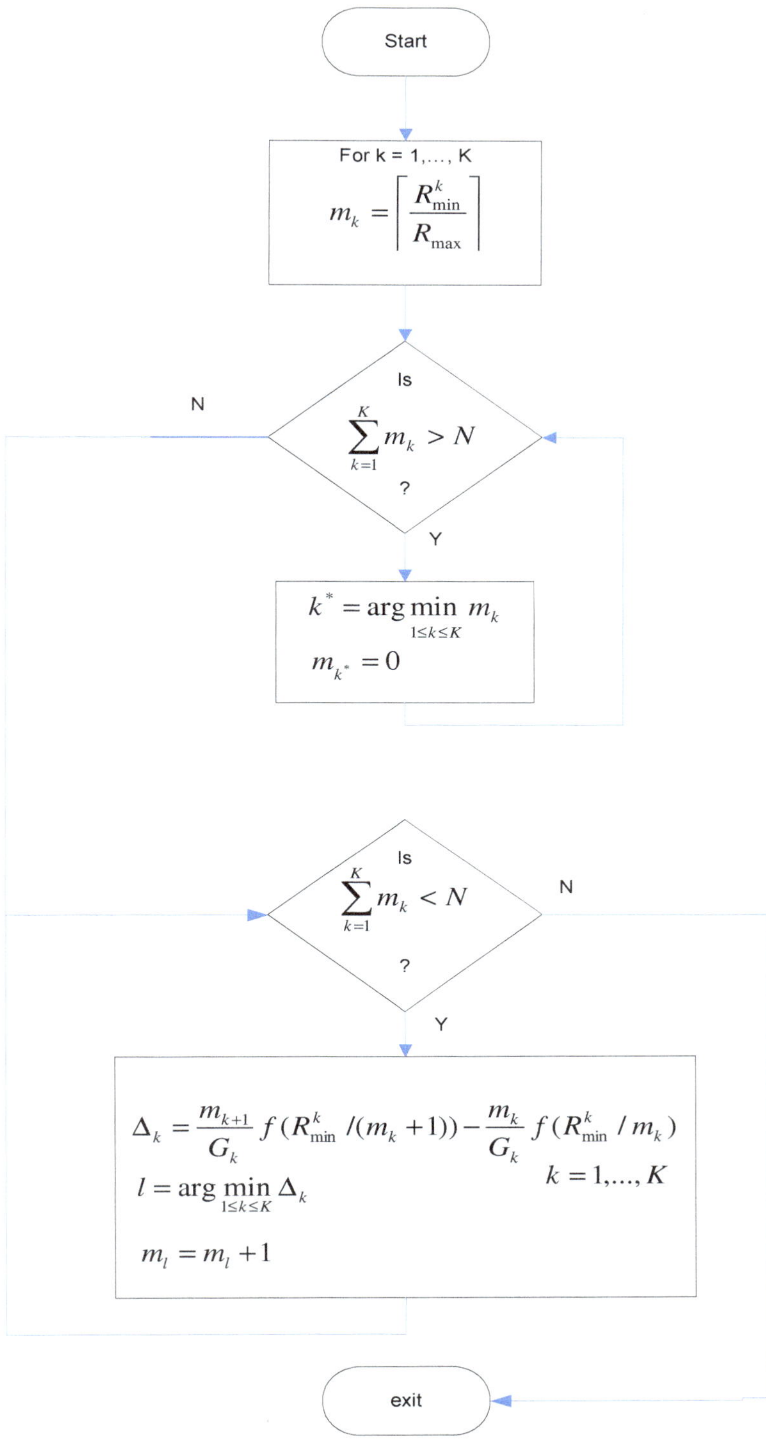

Figure 3 Subcarrier allocation (constructive initial allocation)

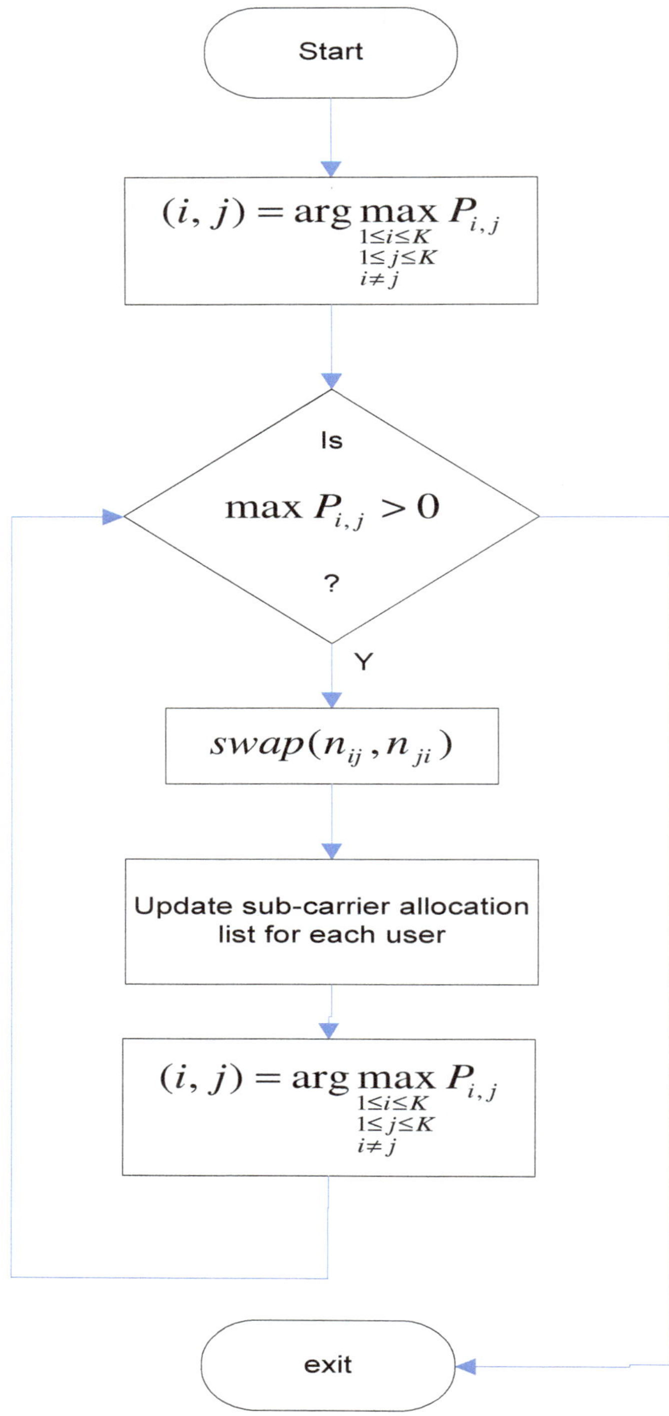

Figure 4 Subcarrier allocation(Iterative improvement)

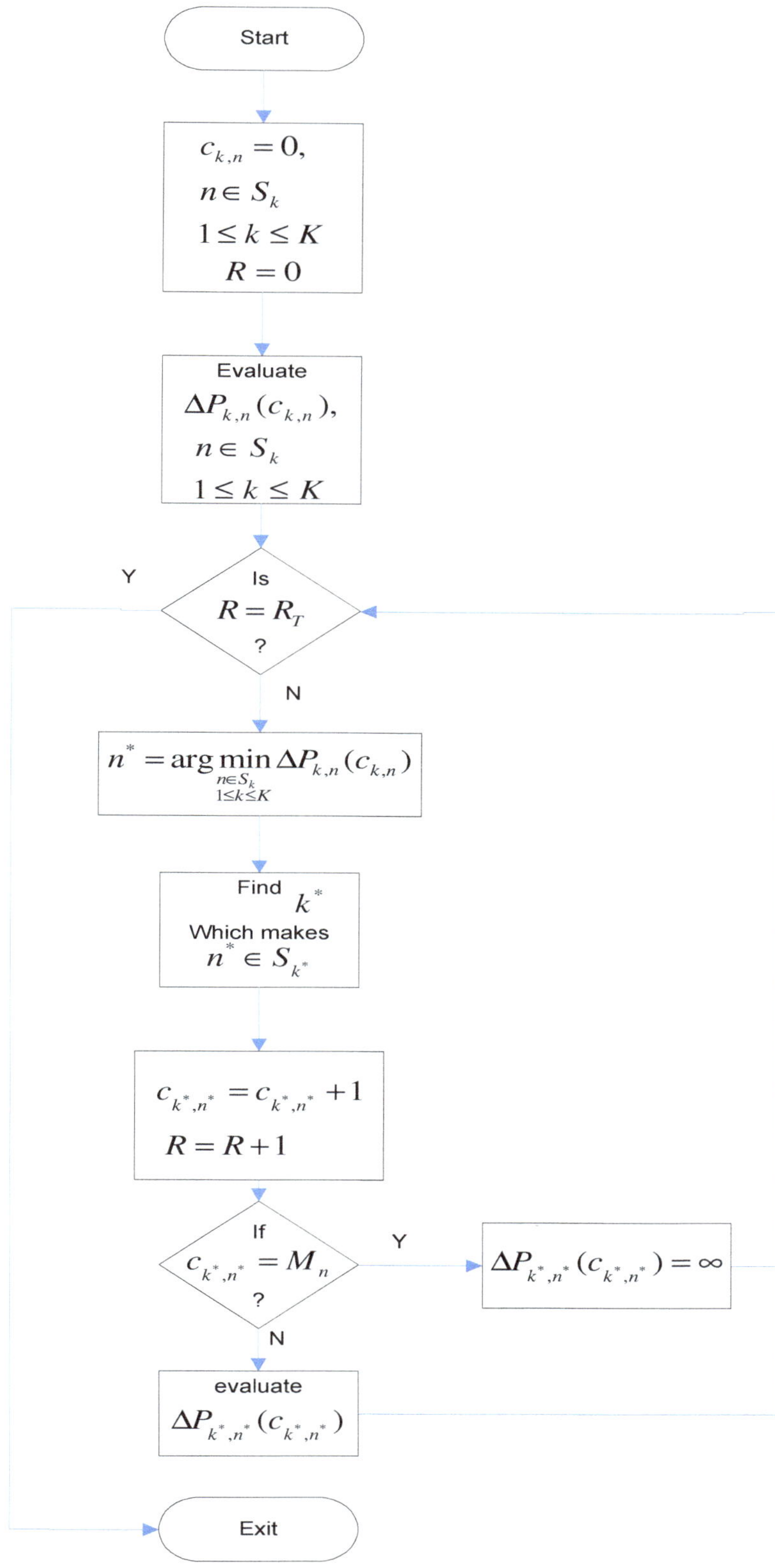

Figure 5 Bit loading

6.4 Primitive parameters

The following four primitive parameters characterize the OFDMA symbol:

BW: This is the nominal channel bandwidth.
*N*used: Number of used subcarriers (which includes the DC subcarrier).
n:Sampling factor. This parameter, in conjunction with *BW* and *N*used determines the subcarrier spacing, and the useful symbol time. This value is set to 8/7.
G:This is the ratio of CP time to "useful" time. The following values shall be supported: 1/32, 1/16, 1/8, and 1/4.

6.5 Derived parameters

The following parameters are defined in terms of the primitive parameters of
Smallest power of two greater than Nused
Sampling Frequency: $Fs = floor(n \,.\, BW/8000) \times 8000$
Subcarrier spacing: $\Delta f = Fs/\,NFFT$
Useful symbol time: $Tb = 1/\Delta f$
CP Time: $Tg = G \,.\, Tb$
OFDMA Symbol Time: $Ts = Tb + Tg$
Sampling time: $Tb/NFFT$

7. RESULTS AND DISCUSSION

Figure 6 shows the results which are calculated for 64 subcarriers and subchannel SNR = 38 dB and SNR gap ¡ = 3:3.The capacity achieved by the proposed LINEAR method is consistently higher than for the NONLINEAR method for a system with 4 up to 16 users. Figure 6 shows the comparison of total capacities between the proposed LINEAR method and NONLINEAR method . Notice that the capacities increase as the number of users increases. This is the effect of multiuser diversity gain, which is more prominent in systems with larger number of users. The proposed LINEAR method has a consistently higher total capacity than the NON- LINEAR method for all the numbers of users for this set. This analysis is done in Matlab R © 2008a

Figure 7 gives the Normalized capacity ratios per user for 16 users averaged over 100 channels, with the required Gamma shown as the leftmost bar for each user. The LINEAR method has minimal deviation from the required proportions, but Proposed scheme is much better for large number of users .The normalized proportions of the capacities for each user for the case of 16 users averaged over 100 channel samples

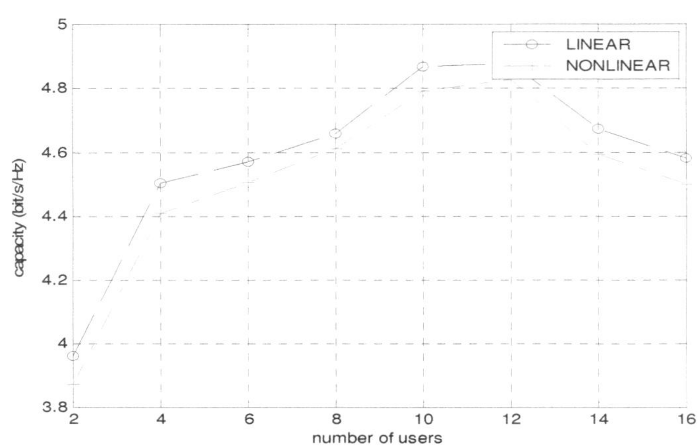

Figure 6 Total capacity versus number of users in an OFDMA system

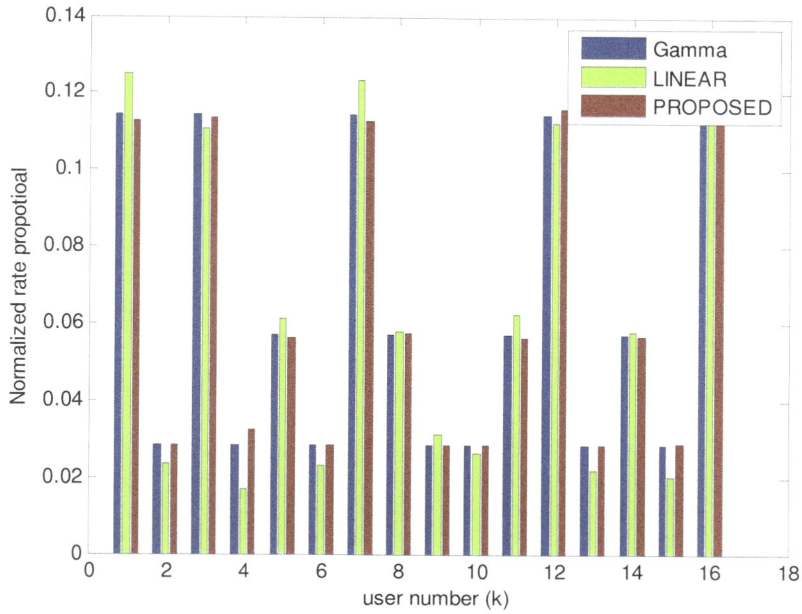

Figure 7 Normalized capacity ratios per user

8. CONCLSION

This paper presents a new method to solve the rate-adaptive resource allocation problem with proportional rate constraints for OFDMA systems. It improves on the previous work in this area by developing a novel subcarrier allocation scheme that achieves approximate rate proportionality while maximizing the total capacity. It is shown through simulation that the proposed method performs better than the previous work in terms of significantly decreasing the computational complexity, and yet achieving higher total capacities, while being applicable to a more general class of systems.

REFERENCES

[1] P. Parag, S. Bhashyam, and R. Aravind, "A subcarrier allocation algorithm for OFDMA using buffer and channel state information," in Proc. 62nd IEEE Veh. Technol. Conf., Sep. 2005, vol. 1, pp. 622–625.

[2] G. Song, Y. Li, L. J. Cimini, and H. Zheng, "Joint channel-aware and queue-aware data scheduling in multiple shared wireless channels," in Proc. IEEE WCNC, Mar. 2004, vol. 3, pp. 1939–1944

[3] Zukang Shen, J. G. Andrews, and B. L. Evans, "Adaptive resource allocation in multiuser OFDM systems with proportional rate constraints," *IEEE Transactions on Wireless Communications*, Vol. 4, pp. 2726 2737, 2005.

[4] I. C. Wong, Z. Shen, J. G. Andrews, and B. L. Evans, "A low complexity algorithm for proportional resource allocation in OFDMA systems," *Proc. IEEE Int. Work. signal Processing Systems* , pp. 1-6, 2004.

[5] Didem Kivanc, Guoqin g Li, and H ui Liu, "Comput ationally Efficient B and width Allocation and Power Control for O FDMA," *IEEE Trans. Wireless Communication.*, vol. 2, pp. 1150-1158, Nov. 2003.

[6] Must afa Er gen, S inem C oleri, and Pravin Varaiya , "QoS Awar e Adapti ve R esource Allocation Techniques for Fair Scheduling in OFDMA Based Broadband Wireless Access S ystems," *IEEE Trans. Broadcasting*, vol. 49, pp. 362-370, Dec. 2003.

[7] Tolga Girici, Chenxi Zhu, Jonathan R. Agre, and Anthony Ephremides,"Proportional Fair Scheduling Algorithm in OFDMA Based Wireless Systems with QoS Constraints" journal of communications and networks, vol. 12, no. 1, February 2010

[8] Alessandro Biagioni, Romano Fantacci, Dania Marabissi and Daniele Tarchi, "Adaptive Subcarrier Allocation Schemes for Wireless OFDMA Systems in WiMAX Networks". IEEE Journal on selected areas in communications, vol.27, No.2, FEB – 2009.

[9] Kibeom S eong, M ehdi M ohseni, and J ohn M. C ioffi, "Optim al R esource Allocati on for OFDMA Downli nk S ystems," *Proc. IEEE Int'l S ymp. on Info. T heor y (ISIT)* , pp. 1394-1398, July 2006.

[10] Mohmmad Anas, Kanghee Kim, Seokjoo Shin, and Kiseon Kim, "QoS aware power allocation for combined guaranteed performance and best effort users in OFDMA systems," *IEEE International Symposium on Intelligent Signal Processing and Communication Systems* ,pp. 477-481, 2004.

[11]. Jiho Jang, Kwang Bok Lee, and Yong-Hwan Lee, "Transmit power and bit allocations for OFDM systems in a fading channel," in *Proc. IEEE Global Communications Conference,* Vol. 2, pp. 858-862, 2003.

[12] H. Yin and H. Liu, An Ef_cient Multiuser Loading Algorithm for OFDM-based Broadband Wireless Systems,î in *Proc. IEEE Global Telecommunications Conference*, vol. 1, 2000, pp. 103 to 107.

[13] W. Rhee and J. M. Ciof_, Increase in Capacity of Multiuser OFDM System Using Dynamic Subchannel Allocation, in *Proc. IEEE Vehic. Tech. Conf.*, Tokyo, Japan, May 2000, pp. 1085 to 1089.

[14] Z. Shen, J. G. Andrews, and B. L. Evans, Optimal Power Allocation in Multiuser OFDM Systems, in *Proc. IEEE Global Communications Conference*, San Francisco, CA, Dec. 2003, pp. 337 to 341.

[15] S. T. Chung and A. Goldsmith, ìDegrees of Freedom in Adaptive Modulation: A Unied View, *IEEE Trans. Commun.*, vol. 49, no. 9, pp. 1561 to 1571, Sept 2001.

[16] Z. Shen, J. G. Andrews, and B. L. Evans, Adaptive Resource Allocation in Multiuser OFDM Systems with Proportional Fairness, *IEEE Trans. Wireless Commun.*, Nov.2005 issue 6 pp 2726-2737.

[17] Guobin Zhang, Suili Feng, "Subcarrier Allocation Algorithms Based on Graph Coloring in Cognitive Radio NC-OFDM System" South China University of Technology, Guangzhou, China Computer Science and Information Technology (ICCSIT), 2010 3rd IEEE International conference pp 535 to 540

[18] IEEE Standard for Local and metropolitan area networks Part 16: Air Interface for Fixed Broadband Wireless Access Systems 802.16, 2004 pp 1-893.

4

SURVIVABILITY AND RELIABILITY ANALYSIS OF THE TRUSTED LINK STATE ROUTING PROTOCOL FOR WIRELESS AD HOC NETWORKS

Eyad Taqieddin[1], Ann Miller[2], and S. Jagannathan[2]

[1]Department of Network Engineering and Security, Jordan University of Science and Technology, Irbid, Jordan
eyadtaq@just.edu.jo
[2]Department of Electrical and Computer Engineering, Missouri University of Science and Technology, Rolla, MO, USA
milleran@mst.edu, sarangap@mst.edu

ABSTRACT

The nodes in a wireless network are responsible for both sending their traffic as well as relaying the traffic of other nodes in the network. This form of collaboration between the nodes is essential for the proper delivery of data. Without fair participation of all nodes in the routing process, some nodes may lose their energy reserves at a high rate compared to other nodes in the network. However, bandwidth and energy are not the only issues in wireless networks; survivability and reliability are critical as well.

Our focus in this work is on two link state routing protocols; OLSR and TLR. We study the effect of using these protocols on the survivability and the reliability of wireless networks. Both analytical and simulation work show that TLR results in better performance due to the inherent energy aware approach and the traffic partition used to reduce congestions in the network.

KEYWORDS

Reliability, Survivability, Energy, Delay, Wireless Networks.

1. INTRODUCTION

In mobile ad-hoc networks, a set of nodes collaborate with each other to guarantee proper communication between any pair of nodes in the network. This is done with no centralized entities to monitor the operation. Due to the dynamic topology, the links in the network are established and broken continuously depending on the velocity, direction and transmission range of the nodes. Unlike wired networks, the nodes in an ad hoc network have no fixed infrastructure and are limited in communication range. Intermediate nodes need to forward the packets in the case where the destination node is not within the transmission range of the source.

Efficient routing in ad hoc networks is needed to properly deliver the packets to their intended destinations with minimal delay and to extend the operational lifetime of the network through energy efficient selection of the routes.

Many energy aware routing algorithms were proposed with the aim of improving the survivability of the network by routing packets from their sources to the destinations with the minimal amount of energy. The Minimum Battery Cost Routing (MBCR) is presented in [8]. This algorithm selects the routes with the maximum total capacity. As the name implies, the routes with the least battery cost among all candidate routes are selected. Another algorithm is Simple Energy Aware Dynamic Source Routing (SEADSR) [3]. This is an extension of DSR in which a node, upon receiving the RREQ, incorporates a delay that is proportional to its energy level. This is done because DSR, by design, does not use cost functions in route selection. The authors in [5] present the Optimal Energy-Delay Routing protocol (OEDR), an algorithm based on finding a tradeoff between delay and energy consumption in LSR and compares that to OLSR. Paths are selected based on the least incurred delay and energy consumed. The Power Aware Multi-Access with Signaling (PAMAS) [6] is a protocol that efficiently uses energy by powering off nodes that are not sending or receiving packets. In [1], a detailed analysis of the network behavior both in steady and transient states is given. When the routing nodes move out of the range, then the packets in transit will either be dropped or delayed and this affects the performance of the network. Another analysis that is given in [9] studies the effect of mobility on the availability of the paths. Both [1] and [9] use a the Continuous Time Markov Model (CTMC) to model the connection availability for a two hop ad hoc network with respect to node mobility and failures.

In [7], we proposed a Link State Routing protocol based on the Optimized Link State Routing (OLSR) [4]. Our proposed protocol, the Trust Levels Routing protocol (TLR), adds security measures to counter the vulnerabilities in OLSR and to better consume the energy resources of the nodes in the network. In this work, we will analyze both OLSR and TLR in terms of survivability and reliability to emphasize the improved performance of the network in terms of lifetime and proper delivery of data. Our focus is to explain analytically the achieved improvements in the energy consumption and its effect on the survivability and reliability. The analysis is further supported by simulation scenarios.

The paper is organized as follows, in section 2 we present an overview of OLSR and TLR. Sections 3 gives a discussion of the survivability analysis of TLR followed by the reliability analysis in section 4. Simulation results are given in section 5 and the paper is concluded in section 6.

2. OVERVIEW OF OLSR AND TLR

The routing protocol should focus on energy conservation to increase the lifetime of the nodes while choosing routes with the least delay, jitter and congestion. Occasionally, the best routes for several sources, in terms of delay, go through the same node whose energy gets consumed at a higher rate compared to other nodes. This, eventually, leads to a premature loss of the battery of the node. A more efficient approach is to route packets through paths that may have higher delays but with more energy resources in order to extend the lifetime of the network.

OLSR and TLR are proactive link state routing protocols. Their operation is table driven through periodically exchanging topology information with other nodes in the network. The cornerstone in the operation of both protocols is the use of Multi Point Relay (MPR) nodes which limit the flooding in the network to a specific set of nodes that are chosen according to the dynamics of the network. In OLSR, the algorithm searches for the smallest set of MPR nodes, i.e. it selects the nodes that cover the largest number of two hop neighbors. As for TLR, the MPR set is selected based on a tradeoff between the delay incurred, energy consumed and load balancing as will be explained later. In addition to energy conservation, TLR provides guarantees of the integrity, timeliness and authenticity of the packets. Our proposed protocol uses the following metrics when considering the selection of the MPR nodes.

2.1. Energy consumed per packet

For this metric, the best path is selected based on the least consumed total energy. Any packet going from source n_1 to destination n_k through some intermediate nodes will consume

$$E_t = \sum_{i=1}^{k-1} (e^t_{(i,\, i+1)} + e^r_{(i,\, i+1)}) \tag{1}$$

where $\mathbf{E_t}$ is the total energy consumed and $\mathbf{e^t}(i, i+1)$ is the energy consumed to send the packet from n_i to n_{i+1} and $\mathbf{e^r}(i, i+1)$ is the energy consumed by n_{i+1} when receiving a packet from n_i.

2.2. Delay per packet

Similar to the energy metric, the goal is to find the path with the least total delay. For a packet going from node n_1 to node n_k through some intermediate nodes, the total delay incurred is given by

$$D_t = \sum_{i=1}^{k-1} d_{(i,\, i+1)} \tag{2}$$

where $\mathbf{D_t}$ is the total delay and d(i, i+1) is the time that starts when a packet enters the queue of node n_i until it reaches the queue of n_{i+1}. Note that a trade off between the two metrics exists.

2.3. Residual energy levels

This metric is used to guarantee that all nodes will have approximately equal rates of consumption by using the nodes with the highest energy levels. That is, the lower the remaining available energy in a node, the higher the cost of routing through it. The cost in this case can be taken as the reciprocal of the residual energy.

2.4. Traffic partition

This metric serves multiple purposes. First, less congestion and delay will be incurred through the intermediate nodes. Second, a higher throughput will be achieved because data packets are going through different paths. More importantly, the traffic will be fragmented into multiple paths which can potentially reduce the ability of a malicious node to capture the whole stream of traffic.

Since TLR bases the selection of MPRs on a composite metric that differs from that of OLSR, the MPR set chosen does not necessarily have to be identical for the same network topology and conditions. Moreover, if the nodes are stationary, the MPR set for nodes running OLSR will always be the same until one or more nodes lose their battery power. In TLR, the MPR set is dynamically changed more frequently based on the MPR routing criteria.

TLR is enforced with security mechanisms to overcome some of the security threats that may impede the operation of the network. The control packets are signed and the data stream is encrypted with using a lightweight encryption algorithm.

2.5. MPR Selction Algorithm

The following notation will be used

N: set of nodes in network

s: source node

d: destination node

N_1(s): One-hop neighbors of node s

$N_2(s)$: Two-hop neighbors of node s

MPR(s): The set of nodes selected as MPRs by node s ($MPR(s) \in N_1(s)$)

$P_{x,y}$: Number of packets sent from *x* through MPR *y*.

w_1 and w_2: Weight factors

$C_{x,y}$: cost of link between nodes x and y, where

$$C_{xy} = w_1 * (Energy_{x \to y} * Delay_{x \to y}) + w_2 * P_{xy} \quad (3)$$

Initially the MPR set MPR(s) is empty.

1. First, find all the nodes in $N_2(s)$ that have a single neighbor in $N_1(s)$. Add these nodes of $N_1(s)$ to the MPR set if they are not already in MPR(s). (Because there are no other MPR candidates).
2. While there exists a node in $N_2(s)$ for which MPR node is not selected, do:
   a. **TLR:** for each node in $N_2(s)$, with multiple neighbors from $N_1(s)$, select a neighbor from $N_1(s)$ as an MPR node which results in minimum cost from s to the node in $N_2(s)$, C_{MPR} according to (3), and add it to the MPR set if is not already in MPR(s)
   b. **OLSR:** for each node in $N_2(s)$, with multiple neighbors from $N_1(s)$, select a neighbor from $N_1(s)$ as an MPR node which covers the maximum number of two hop neighbors in $N_2(s)$ and add it to the MPR set if it is not already in MPR(s).

3. SURVIVABILITY ANALYSIS OF TLR

The survivability of the network is directly related to the lifetime of nodes in the network. Clearly, the energy in every node in the network has to be properly utilized in order to avoid premature depletion of energy of the nodes and to minimize the variance in energy levels between them. The power consumption is due to 1) transmission of a packet; 2) reception of a packet 3) retransmission of packets due to congestion 4) power used when a node is idle.

Given a network modelled as a graph G = (N,L) where N is the set of nodes and L is the set of links (i,j) ϵ N. A link (i,j) exists if and only if j ϵ $N_1(i)$. The lifetime of a node in the network running the OLSR protocol is given as

$$LT_i = \frac{Ei}{(1+\gamma)\sum_{k \in N1(i)} e_{tx} r_{i \to k} + \sum_{j:i \in N1(j)} e_{rx} r_{j \to i} + \varepsilon} \quad (4)$$

where,

Ei: Energy level of node i at t = 0 (Joule)

e_{tx}: Energy consumed to transmit data (Joule/bit)

e_{rx}: Energy consumed to receive data (Joule/bit)

$r_{i \to j}$: Data rate between source i and destination j (bit/sec) ε: Energy consumed during the idle time (Joule/sec)

γ: Factor representing the needed retransmissions

As for a node running TLR, the lifetime is

$$LT_i = \frac{Ei}{(\alpha+\beta\gamma)\sum_{k\in N1(i)} e_{tx} r_{i\to k} + \sum_{j:i\in N1(j)} e_{rx} r_{j\to i} + \varepsilon + \mu} \quad (5)$$

with

μ: Energy used for overhead operations in TLR (encryption functionality).

α and β: Factors comprising the overall number of packets transmitted.

The total energy used for the transmission of packets and the total energy to compensate for repeated packets due to congestion are multiplied by factors α and β (both less than 1), respectively. This is a result of the load balancing in TLR which results in a node being used less frequently compared to the case of OLSR. Moreover, load balancing results in less congestion leading to fewer dropped packets. Note that the factors α and β are inversely proportional to the number of available MPR candidates.

TLR outperforms OLSR in terms of energy conservation by distributing the load on various intermediate nodes. This leads to a relatively smaller variance in the battery levels and longer lifetime for the whole network.

To illustrate, consider the network given in Figure 1 where nodes S1 and S2 send traffic through nodes A, B and C to the destinations D1 and D2, respectively. Assuming that each of nodes S1 and S2 has 50 packets to send and also assume that each of nodes A, B and C has sufficient energy to receive and forward 100 packets only. With OLSR, the node that covers the maximum number of hops will be selected as the MPR. In this case, node C will be selected because it covers 2 nodes (D1 and D2) whereas nodes A and B only cover D1 and D2, respectively. Thus, both nodes S1 and S2 will send all their traffic through C which will result in consuming all it energy resources after sending 50 each (100 in total). After that, S1 will select A as the MPR whereas S2 will select B. In this scenario, node C lost all its energy prematurely whereas nodes A and B still maintained their initial energy reserves. If either node A or B moves out range then network will be partitioned due to the lack of available paths.

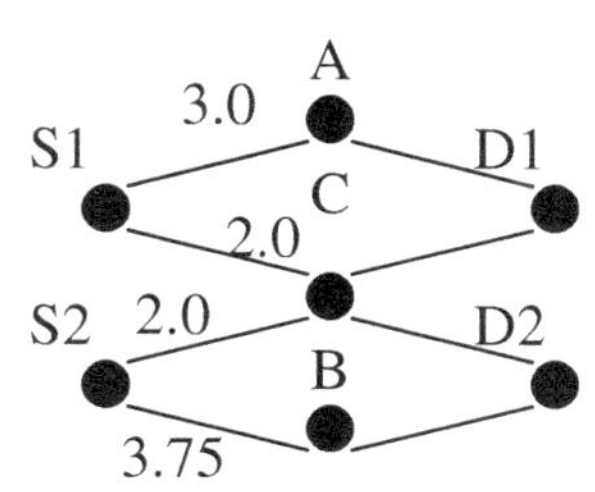

Figure 1. Ad hoc network with seven nodes.

On the other hand, when TLR is used with the same network topology using the initial link costs given in figure 1, node C will be initially selected as the MPR because it has the least cost for both S1 and S2. After forwarding 20 packets, the link cost for S1-C and S2-C will be 3.5. Thus, on the next topology update, S1 will switch to node A for forwarding whereas S2 will maintain the same link S2-C. After transmitting 20 more packets, the updated link costs will be (S1-A = 4.33, S1-C = 2.86, S2-C =3.86, S2-B = 3.75. As a result, node S1 will use the link S1-C and node S2 will use S2-B. Table 1 details the energy levels of the nodes throughout the above mentioned scenario. Note that node C still retains 40% of its energy whereas it consumed all its energy when OLSR was used.

Table 1. Remaining energy levels of the three routing nodes using TLR

# Packets Sent	Link Costs				Active Link		Remaining Energy		
	S1-A	S1-C	S2-C	S2-B	S1	S2	A	B	C
0	3.0	2.0	2.0	3.75	S1-C	S2-C	100%	100%	100%
20	3.0	3.5	3.5	3.75	S1-A	S2-C	100%	100%	80%
40	4.33	2.86	3.86	3.75	S1-C	S2-B	90%	100%	70%
60	3.33	4.333	3.33	5.17	S1-A	S2-C	90%	90%	60%
80	4.75	4.0	5.0	4.17	S1-C	S2-B	80%	90%	50%
100	3.75	6.0	5.0	5.69	S1-A	S2-C	80%	80%	40%

4. RELIABILITY ANALYSIS OF TLR

The reliability of a system is defined as the conditional probability that the system is operational at time t given that it was operational at time t_0. Similarly, the reliability of a routing protocol can be given as the conditional probability that the network will properly deliver the packets from the source to the destination given that the routing paths were available at time t_0.

Considering that a network is a system consisting of nodes and links then the reliability of the network will be directly dependent on the reliability of the components. Any failure to either of the nodes or links may lead to disruption of the service of the network. The service may remain available due to the redundancy of paths through various nodes. In this section, we analyze the reliability of OLSR and TLR for stationary nodes using Markov analysis.

4.1. Markov Analysis

Markov analysis provides a way of studying the sequence of events that a system may execute. This sequence of random and related events is useful in determining the reliability and availability of the system since it reflects the transition probabilities between the system states.

A Markov model is a function of two random variables; the time of observation and the current state, irrespective of the previous sequence that led to the current state.

The main advantages using Markov analysis are the simplicity of constructing the model, availability of transition probabilities between states and the incorporation of component redundancy. However, one drawback of this form of analysis is that the complexity increases with the number of states.

A Continuous Time Markov Chain (CTMC) can be used to model a system with a finite number of states to obtain a closed form solution of the reliability or availability of the system. For simplicity, we will assume that 1) component failures are mutually independent; 2) failures occur at an exponential rate; 3) equal repair rate for all component failures and 4) only one failure can occur during transition between states. The last condition is to guarantee that no two routers fail during a switching delay period.

Consider the Markov reliability model given in figure 3. The probability of transition from state S_0 to S_1 in time Δt is equal to $\lambda_0 \Delta t$ and the probability of remaining the same state is $(1- \lambda 0)\ \Delta t$. We can find the probabilities of the system being in either of the states at time $t = t+ \Delta t$ using the difference equations.

$$P_{S0}(t+\Delta t)=(1-\lambda_0 \Delta t)P_{S0}(t) \tag{6}$$

$$P_{S1}(t+\Delta t)=(\lambda_0 \Delta t)P_{S0}(t)+(1-\lambda_1 \Delta t)P_{S1}(t) \tag{7}$$

$$P_{S2}(t+\Delta t)=(\lambda_1 \Delta t)P_{S1}(t)+(1-\lambda_2 \Delta t)P_{S2}(t) \tag{8}$$

$$P_{S3}(t+\Delta t)=(\lambda_2 \Delta t)P_{S2}(t)+P_{S3}(t) \tag{9}$$

Rearranging the equations:

$$\frac{P_{S0}(t+\Delta t)-P_{S0}(t)}{\Delta t}=-\lambda_0 P_{S0}(t) \tag{10}$$

$$\frac{P_{S1}(t+\Delta t)-P_{S1}(t)}{\Delta t}=\lambda_0 P_{S0}(t)-\lambda_1 P_{S1}(t) \tag{11}$$

$$\frac{P_{S2}(t+\Delta t)-P_{S2}(t)}{\Delta t}=\lambda_1 P_{S1}(t)-\lambda_2 P_{S2}(t) \tag{12}$$

$$\frac{P_{S3}(t+\Delta t)-P_{S3}(t)}{\Delta t}=\lambda_2 P_{S2}(t) \tag{13}$$

When $\Delta t \to 0$

$$\frac{dP_{S3}(t)}{dt}=\lambda_2 P_{S2}(t) \tag{14}$$

$$\frac{dP_{S2}(t)}{dt}=\lambda_1 P_{S1}(t)-\lambda_2 P_{S2}(t) \tag{15}$$

$$\frac{dP_{S1}(t)}{dt}=\lambda_0 P_{S0}(t)-\lambda_1 P_{S1}(t) \tag{16}$$

$$\frac{dP_{S0}(t)}{dt}=-\lambda_0 P_{S0}(t) \tag{17}$$

The set of differential equations (14)-(17) can be solved to find the overall reliability as will be shown in section 4.2.

The analysis will use the CTMC to find the reliability of a network consisting of stationary nodes with no repair of dead nodes.

4.2. Reliability Analysis

The reliability of any ad hoc network depends on the reliability of the nodes. Thus, the failure of the network will be proportional to the expected lifetime of the nodes in the network. The network reliability and survivability calculation, however, is not simple; at some point, given the topology and the number/location of failed nodes, a network will fail. This will be case/state specific. However, we can make the following generalizations. The failure rate, λ, is defined as the number of failures per unit of time. One of the advantages of TLR is that it increases the average lifetime of all the nodes and decreases the variance of energy levels between the nodes compared to OLSR. In effect, TLR decreases the failure rate of the nodes which results in higher reliability.

By design, OLSR runs the nodes in a standby mode where only one node is responsible of routing and all other nodes remain in idle mode. When the routing node dies, another node will be selected to perform the routing. On the other hand, TLR runs the nodes in a parallel mode where one node handles the routing functionality for some time then another node takes over. The routing nodes in such a scenario will have a longer lifetime as was shown in (5).

Consider the network given in figure 2, in which S1 has no direct access to its two hop neighbor D1. Instead, nodes A, B and C will be used to route the traffic.

For OLSR, the reliability Markov model is given in figure 3. The system may be in one of the four states {Si I i = 0,1,2,3}. State S_0 represents the initial state of the network at time t = 0 and all three nodes will be functional whereas S_3 is the state when all three routing nodes fail.

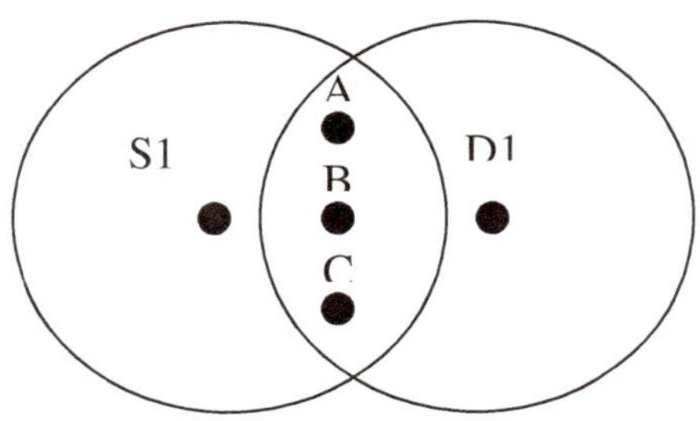

Figure 2. Ad hoc network with five nodes.

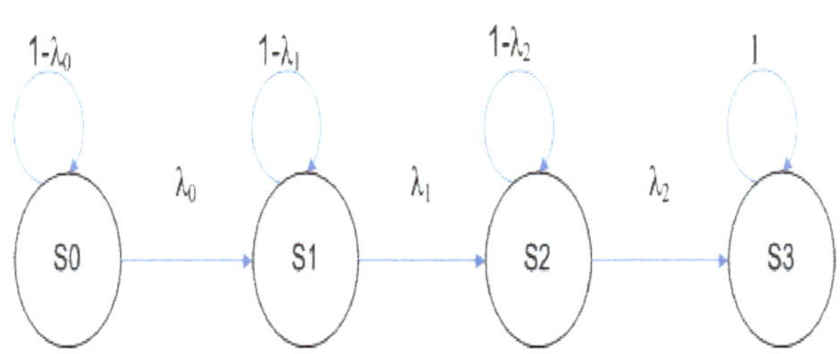

Figure 3. Markov Reliability Model for the network running OLSR

The probabilities of each of the states can be calculated using equations. (14) –(17) with the initial conditions $P_{S0}(0) = 1$, $P_{S1}(0) = P_{S2}(0) = P_{S3}(0) = 0$. Taking the Laplace transform of the above equations yields

$$sP_{S3}(s)=\lambda_2 P_{S2}(s) \tag{18}$$

$$sP_{S2}(s)=\lambda_1 P_{S1}(s)-\lambda_2 P_{S2}(s) \tag{19}$$

$$sP_{S1}(s)=\lambda_0 P_{S0}(s)-\lambda_1 P_{S1}(s) \tag{20}$$

$$sP_{S0}(s)-1=-\lambda_0 P_{S0}(s) \tag{21}$$

Rearranging the equations

$$P_{S0}(s)=\frac{1}{s+\lambda_0} \tag{22}$$

$$P_{S1}(s)=\frac{\lambda_0}{(s+\lambda_0)(s+\lambda_1)} \tag{23}$$

$$P_{S2}(s)=\frac{\lambda_0\lambda_1}{(s+\lambda_0)(s+\lambda_1)(s+\lambda_2)} \tag{24}$$

Converting back to the time domain

$$P_{S0}(t)=e^{-\lambda_0 t} \tag{25}$$

$$P_{S1}(t)=\frac{\lambda_0}{(\lambda_1-\lambda_0)}\left[e^{-\lambda_0 t}-e^{-\lambda_1 t}\right] \tag{26}$$

$$P_{S2}(t)=\lambda_0\lambda_1\left[\frac{e^{-\lambda_0 t}}{(\lambda_1-\lambda_0)(\lambda_2-\lambda_0)}+\frac{e^{-\lambda_1 t}}{(\lambda_0-\lambda_1)(\lambda_2-\lambda_1)}+\frac{e^{-\lambda_2 t}}{(\lambda_0-\lambda_2)(\lambda_1-\lambda_2)}\right] \tag{27}$$

If $\lambda_0=\lambda_1=\lambda_2=\lambda$ then we can find $P_{S1}(t)$ and $P_{S2}(t)$ using L'hospital's rule (ordinary substitution would not work because the equations would yield 0/0).

$$P_{S0}(t)=e^{-\lambda t} \tag{28}$$

$$P_{S1}(t)=\lambda t e^{-\lambda t} \tag{29}$$

$$P_{S2}(t)=\frac{(\lambda t)^2}{2}e^{-\lambda t} \tag{30}$$

The network will remain operational as long as it remains in S_0, S_1, or S_2. Thus, the reliability of the network will be

$$R(t)=P_{S0}(t)+P_{S1}(t)+P_{S2}(t)=e^{-\lambda t}+\lambda t e^{-\lambda t}+\frac{(\lambda t)^2}{2}e^{-\lambda t} \tag{31}$$

An interesting observation is that the terms in R(t) are the first three terms in the Poisson distribution. It can be read as the probability of zero failures in time t plus the probability of one failure in time t plus the probability of 2 failures in time t. Extending the model to n nodes would result in

$$R(t)=\sum_{i=0}^{n-1}\frac{(\lambda t)^i}{i!}e^{-\lambda t} \tag{32}$$

Figure 4 shows the reliability Markov model for the TLR protocol. Here, the transition probability from state S_0 to S_1 is $3\lambda\Delta t$ because all three nodes are functional and anyone of them could fail. The same logic can be applied to the other transitions in the figure.

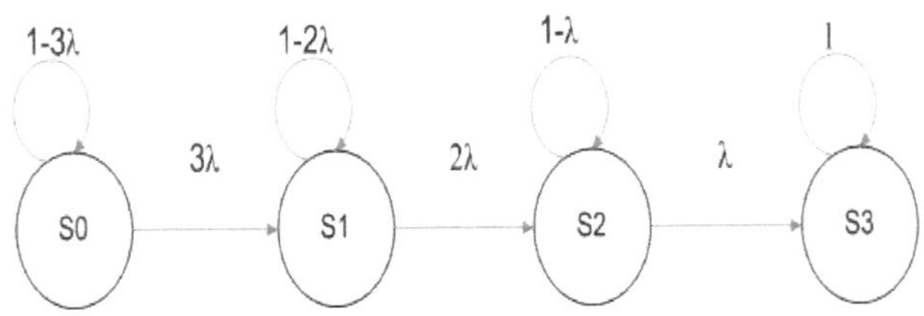

Figure 4. Markov Reliability Model for the network running TLR

The reliability can be either calculated in a similar manner to that done for OLSR or we can use the fact that all the components are working in parallel and the system will be functional if either one of them is working.

$$R(t)=P(A+B+C)=3e^{-\lambda t}-3e^{-2\lambda t}+e^{-3\lambda t} \tag{33}$$

The reliability of a network with n routing nodes can be given as

$$\begin{aligned} R(t) = & \sum_{i=1}^{n} P(A_i) - \sum_{i=1}^{n} \sum_{j=i+1}^{n} P(A_i A_j) \\ & + \sum_{i=1}^{n} \sum_{j=i+1}^{n} \sum_{k=j+1}^{n} P(A_i A_j A_k) - \ldots \\ & + (-1)^{n-1} [P(A_1 A_2 \ldots A_n)] \end{aligned} \tag{34}$$

Looking back at (5), we see the lifetime of a node will increase as the factors α and β decrease. Evidently, with more MPR candidate these factors must decrease since the node will be forwarding or retransmitting fewer packets. When only one MPR is available then α and β will be equal to 1.0 as the packets must follow the same path.

The relation between the failure rates for nodes running TLR (λ_{TLR}) and that of OLSR (λ_{OLSR})is

$$\lambda_{TLR} = \left[\frac{(\alpha+\beta\gamma) \sum_{k\in N1(i)} e_{tx} r_{i->k} + \sum_{j:i\in N1(j)} e_{rx} r_{j->i} + \varepsilon + \mu}{(1+\gamma) \sum_{k\in N1(i)} e_{tx} r_{i->k} + \sum_{j:i\in N1(j)} e_{rx} r_{j->i} + \varepsilon} \right] \lambda_{OLSR} \tag{35}$$

To illustrate, assume that the failure rate of nodes resulting from running the OLSR and TLR protocol are λ_{OLSR}=0.05 and λ_{TLR}=0.025, respectively. The reliability functions of the network for both protocols are given in figure 5.

If the parameters of the network such as the congestion and data rates are known then we can determine the suitable encryption algorithm to be used based on the value of μ, according to the following condition

$$\mu < ((1-\alpha)+\gamma(1-\beta)) \sum_{k \in N1(i)} e_{tx} r_{i->k} \tag{36}$$

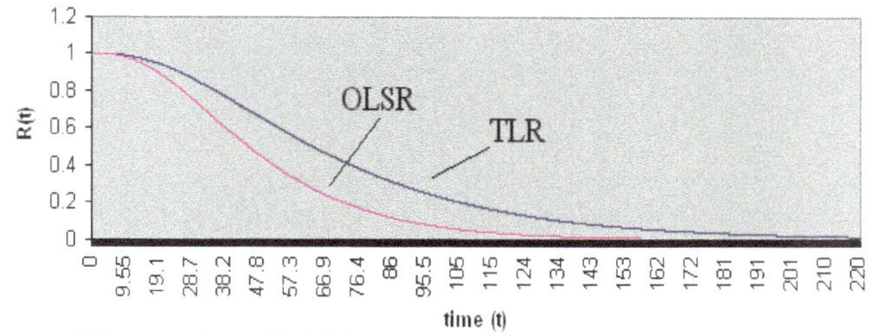

Figure 5. Reliability curves for OLSR and TLR

5. SIMULATION RESULTS

We used the network simulator NS2 to study the effect of running TLR on the survivability of the nodes in the network. We implemented the algorithm as an extension to the existing OLSR code by modifying the selection of the MPR nodes in NS2. The energy values for transmission were increased to simulate the extra energy needed for encryption. The simulations were conducted in a 670 x 670 area with 100 nodes. The nodes communicate using Constant Bit Rate (CBR) sources that are connected to the nodes in a random fashion. The data rates of 64, 128, 256 and 512 kpbs were used and the results of the runs were averaged. Each run had a period of 100 seconds.

The first objective of the simulation was to show the distribution of the energy in the nodes after running for the whole simulation period. After each run, the remaining energy of each node was sorted into one of 10 categories based on the percentage of the remaining energy compared to the energy level at the beginning of the simulation.

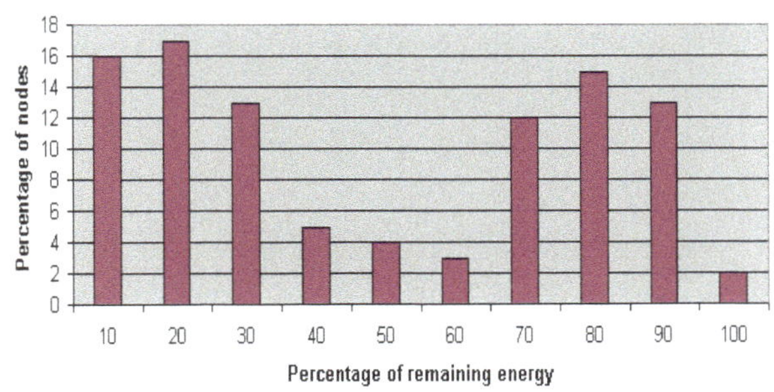

Figure 6. Energy distribution in nodes running OLSR

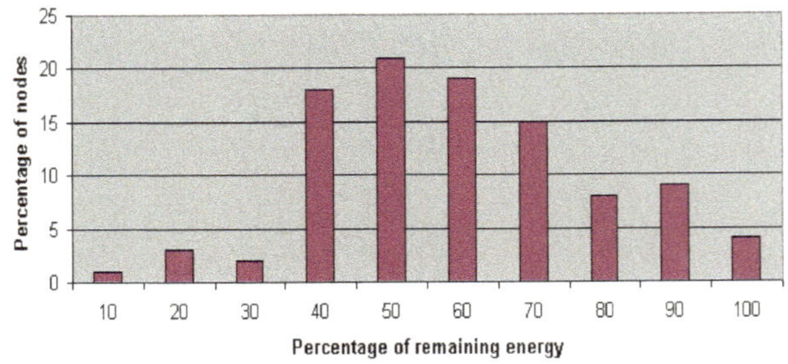

Figure 7. Energy distribution in nodes running TLR

We note from figure 6 that the percentage of remaining energy in the nodes running OLSR is not distributed optimally throughout the network. On one hand, we notice that a high percentage of the nodes hold a high percentage of their initial energy which means they were not actively involved in the routing process. This might seem to be a positive feature but it comes at a cost. Namely, another high percentage of nodes have little remaining energy. Only a few nodes have moderate consumption of their energy during the simulation.

This distribution is not desirable because it shows that around 46% of the nodes had energy levels of 30% or less.

A more desirable distribution of energy is shown in figure 7. In this figure it is evident that a very small percentage of nodes have 30% of less of their initial energy. Also, the percentage of nodes with the higher energy reserves, 80% and above, is smaller than what was shown in the case of OLSR. Most of the energy is distributed among the nodes equally which leads to an extended lifetime of the nodes in the network.

The second objective of the simulation was to study the average energy reserves of all the nodes in the network throughout the period of the simulation. This was done by using the energy levels given in the trace files of NS2. The sampling of the energy was done in 10 second intervals.

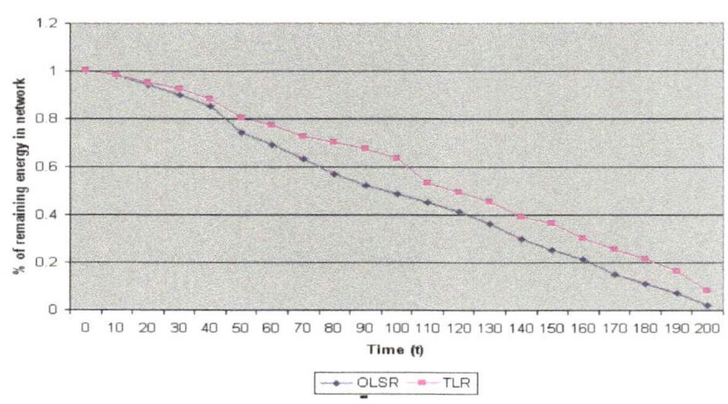

Figure 8. Energy distribution in nodes running OLSR

From figure 8, it is evident that TLR leads to a higher average percentage of remaining energy. This is attributed to the decrease in dropped packets in the network. We see that TLR will give a longer lifetime of the network.

6. Conclusions

Two link state routing protocols were compared in terms of survivability and reliability. Markov models were used to find the reliability function and it was shown how the reliability of TLR is higher than that of OLSR.

The results of this work show the advantage of using TLR instead of OLSR. Notably, TLR results in an extended lifetime of the nodes which may translate to a higher availability of paths.

Future extension of this work includes the analysis of the routing protocols in the case of mobility. In such a case, the intermediate node may move in and out of the communication range.

References

[1] Chen, D., Garg, S., and Trivedi, K. Network Survivability Performance Evaluation: A Quantitative Approach with Applications in Wireless Ad Hoc Networks. In *Proceedings of the Fifth ACM International Workshop on Modeling, Analysis and Simulation of Wireless and Mobile Systems*, Atlanta, USA, 2002

[2] Chen, H., Mineno, H. and Mizuno, T. An Energy-Aware Routing Scheme with Node Relay Willingness in Wireless Sensor Networks. In *Proceedings of the First International Conference on Innovative Computing*, Beijing, China, 2006.

[3] Domingo, M., Remondo, D. and Leon, O. A simple Routing Scheme for Improving Ad Hoc Network Survivability. In *Proceedings of IEEE GLOBECOM'03,* San Francisco, USA, December 1-5, 2003, 718–723.

[4] Jacquet, P. Muhlethaler, P. Clausen, T. Laouiti, A. Qayyum, A., and Viennot, L. Optimized Link State Routing Protocol for Ad Hoc Networks. In *Proceedings of the IEEE International Multi Topic Conference on Technology for the 21st Century, (INMIC'01)*, December, 2001, 62 – 68.

[5] Regatte, N., and Sarangapani, J. Optimized Energy-Delay Routing in Ad Hoc Wireless Networks. In *Proceedings of World Wireless Conference,* San Francisco, USA. *May, 2005.*

[6] Singh, S. and Raghavendra, C.S. PAMAS – Power Aware Multi-Access Protocol with Signaling for Ad Hoc Networks. *ACM Computer Communications Review,28, 3* (July 1998), 5 - 26.

[7] Taqieddin, E., Sarangapani, J. and Miller, A. Optimal Energy-Delay Routing Protocol with Trust levels for Wireless Ad Hoc Networks. In *Proceedings of International Telemetry Conference,(ITC'05)*, Las Vegas, Nevada, Oct 2005.

[8] Toh, C-K, Cobb, H., and Scott, D. Performance Evaluation of Battery-Life-Aware Routing Schemes for Wireless Ad Hoc Networks. In *Proceedings of IEEE International Conference on Communications (IEEE ICC)*, Helsinki, Finland, June 2001.

[9] Trajanov, D., Filiposka, S., Efnuseva, M. and Grnarov, A. Ad Hoc Networks Connection Availability Modeling. In *Proceedings of the 1st International ACM Workshop on Performance Evaluation of Wireless Ad Hoc, Sensor, and Ubiquitous Networks*, Venezia, Italy, 2004, 56 - 60.

[10] Zhou, Y., Laurenson, D., and McLaughlin, S. High Survival Probability Routing in Power-aware Mobile Ad Hoc Networks. *IEE Electronics Letters, Vol. 40, No. 22*, (Oct. 2004), 1424-1426.

5

System Level Simulation for Two Tier Macro-Femto Cellular Networks

[1]Shiqi Xing, [2]Pantha Ghosal, [3]Shouman Barua, [4] Ramprasad Subramanian and [5]Kumbesan Sandrasegaran

Centre for Real-time Information Networks
School of Computing and Communications, Faculty of Engineering and Information Technology, University of Technology Sydney, Sydney, Australia.

Abstract

LTE is an emerging wireless communication technology to provide high- speed data service for the mobile phones and data terminals. To improve indoor coverage and capacity Femtocells are included in 3GPP since Release 8. There is no common simulation platform is available for performance justification of LTE-Femtocells. LTE-Sim is an object-oriented open source simulator which incorporates a complete protocol stack can be used for simulating two-tier macro-femto scenarios. To the best of our knowledge no paper provides the guideline to perform system level simulation of Femtocell networks. Here, in this paper Femtocells performance is evaluated in multi-Macrocells and multi-Femtocells environment with interference from Microcells and Macrocell users along with the scripting.

Keywords

Channel quality indicator (CQI), Femto Access Point (FAP), Macro eNodeB (MeNB), Macrocell User Equepment (MUE), Moblity Management Entity(MME), Signal to Interference Plus Noise Ratio(SINR), Physical Layer(PHY)

1.Introduction

Long Term Evaluation (LTE) is the next generation communication technology. LTE, which supports high data transmission rate has upgraded the major network infrastructure by legacy networks (GSM, GPRS, UMTS/HSPA) with the combination of circuit-switched and packet-switched technology. With the higher frequency bands allocated to LTE, sometimes indoor coverage becomes poor since penetration and path loss is higher for higher frequency bands. Addressing the challenge of indoor coverage, and to provide high data transmission rate to users, small, portable, short ranged (10-35m), low powered (10-100mW) Femto access [1] points were integrated in 3GPP. The benefits of Femtocells include high coverage, high mobile data capacity, improve battery life of mobile devices in the next generation wireless communication system. Femtocells is operating at the frequency spectrum owned by the service operator and enable fixed mobile convergence (FMC) [2] service by connecting to the core network via broadband communication links such as Digital Subscriber Line (DSL). The portability and low cost makes femtocells deployed by the owner in an unplanned manner; also the dense installation of femtocells increases the complexity in mobility management and increases interference. For that purpose research on performance evaluation on accurate simulation framework is of utmost importance. The current open-source simulation tools are only able to simulate parts of the LTE system [3] but we are not aware of any simulator that allows femtocell architecture the media access control (MAC) and physical layer (PHY) models with handover strategies and propagation loss models for indoor environments. To bridge this gap, this paper describes our efforts to build multi-cell and multi-user environment in femtocellular module in the existing open source *LTE-Sim* framework [4] encompassing scenarios with both macro and femto cells

The rest of the paper is organized as, Section II. LTE Network Architecture with Femtocells. Section III. Modelling and Coding in *LTE-Sim.* Section IV explains the simulation outcomes of Femtocells module in *LTE-Sim.* Section V concludes the paper.

2.LTE FEMTOCELL: NETWORK ARCHITECTURE

2.1 LTE and Scheduling Frame

The LTE network architecture is divided into two parts: the air interfaces E-UTRAN, and the packet switched core network (EPC) [5]. On the network side, the evolved NodeB (eNB) is the only component of E-UTRAN, the main functionalities are 1) sending and receiving radio transmission to all mobiles and data terminals in downlink and uplink respectively and 2) controlling singling messages (such as Handover message) On the physical layer, LTE radio interfaces supports both of Frequency Division Duplexing (FDD) and Time Division Duplexing (TDD). Channel is based on Orthogonal Frequency Division Multiple Access (OFDMA). The use of OFDMA removes the Inter Symbol Interferences since the transmitter uses a guard period (cyclic prefix) longer than the channel response [6]. Radio resources are allocated in a time/frequency domain. In the time domain, every Transmission Time Interval (TTI) has been distributed into 1 ms. Each TTI is composed of two consecutive time slots of 0.5 ms and seven OFDM symbols with short cyclic prefix. On the other side, in the frequency domain, the bandwidth is divided into 180 kHz sub-channels, corresponding to 12 consecutive and equally spaced sub-carriers.

2.2 Femtocell and Its challenges

Femtocell was introduced by 3GPP in Release 8 which has the functionalities of boosting the indoor coverage and capacity. Femtocell may also improve battery life of mobile devices. Fig.1 shows the basic femtocell network architecture of macro-femto two-layer systems. According to [3], femtocell access points are connected to the operator network through a DSL line at end user like a common Wi-Fi access point.

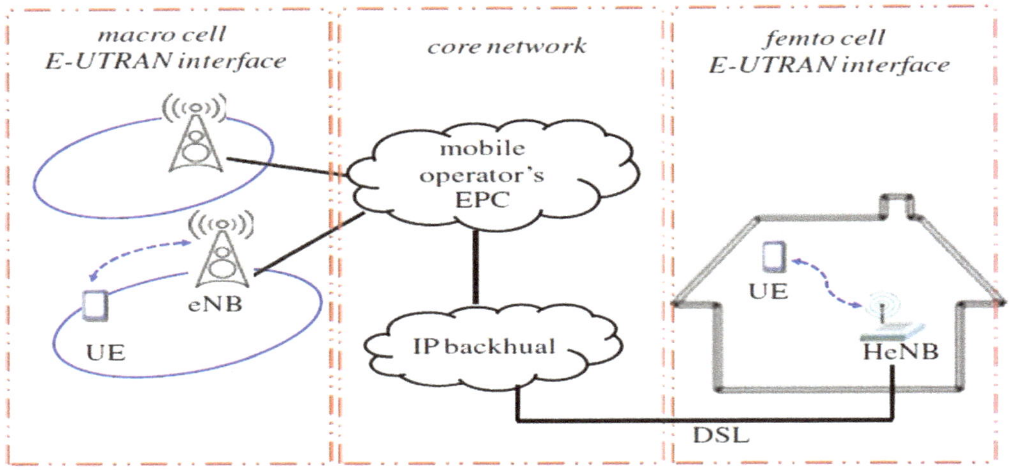

Fig.1. UEs inside building served by Femtocell

LTE system is operating on flat IP architecture which connects all the eNBs through X2 interfaces and Radio Resource Management (RRM) procedure is done by eNB. On the other hand, there is no such X2 interface between eNB and FAP. As a result, Interference coordination [7], RRM and HO management [8] are becoming complex with the unplanned and vast deployment of LTE femtocells. Benefits of femtocells are coming with the increasing the complexity in spectrum management. Femtocells are designed to operate in one of the three different access modes, i.e., closed access mode, open access mode or hybrid access mode [9]. In closed access mode, the UEs belong to the Closed Subscriber Group are allowed to get access to the femtocell. However, the unauthorized users might affect by the interference from the femtocells in closed access mode. In the case of open access mode, any UE could get access to the femtocells and benefit from its services. On the other hand, depending on resource usage, congestion and security consideration, open access is not a suitable choice. For hybrid access mode, a limited number of unknown Macro UEs are allowed to access the femtocell by owners, but the owners will suffer the risk of a security breach [10].

2.3 Femtocell Hand Over Modeling

According to [5], Fig. 2 illustrates three femtocell handover scenarios: Inbound Handover, i.e., the handover that takes place when user moves from the eNB to HeNB. Outbound scenario denotes the handover from the HeNB to eNB. Inter FAP, i.e., UE handover from one femtocell to other femtocells. Fig. 3 shows the Femtocell UE hand over to another femtocell.

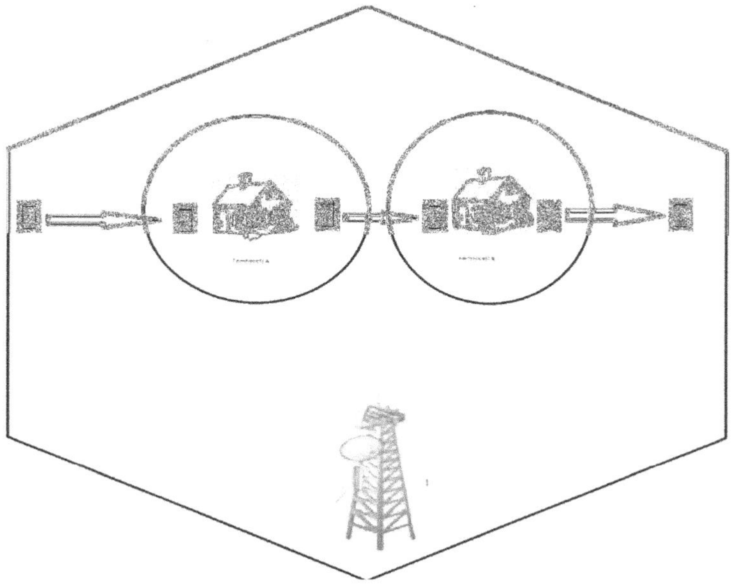

Fig.2: Handover modeling in presence of Femtocells [5]

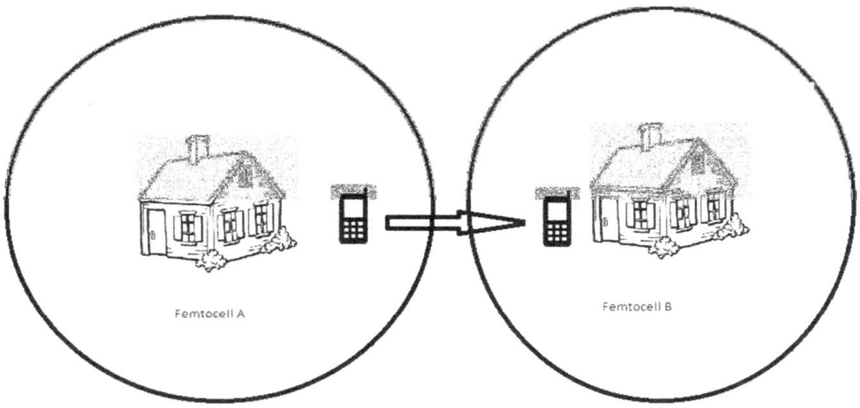

Fig.3: Handover between Femtocells. [5]

2.4 Frequency reuse modeling

LTE supports Fractional Frequency Reuse. The main idea of FFR is to divide allocated spectrum into several sub-bands and allocate these sub-bands in different macro cell coverage area or sub-area [6]. Since, the resources are not overlapped in different sub-areas the cross-tier interference between FAP and eNB can be mitigated. In [11] authors proposed "soft frequency reuse" scheme where frequency spectrum is divided in centre and edge zones by adjusting transmitting power for effective reuse.

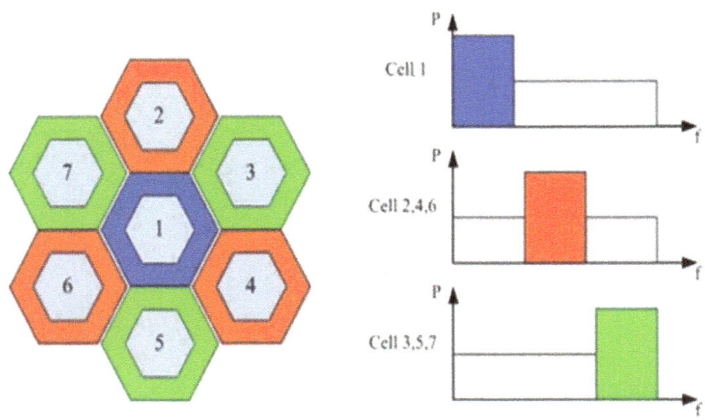

Fig.4: Frequency Planning and Power Allocation for SFR Scheme [11]

In Soft Frequency Reuse, the sub bands are orthogonal to each other and users at cell edge are allocated a cell-edge band as illustrated in Fig. 4 with cluster of 7 cells. Users in the cell edge band have higher power whereas Cell Centre Users of neighbouring cells also have access to the selected cell edge bands but with lower power. Therefore, high spectral efficiency could be achieved with user throughput. Though, intracell interference is reduced because of orthogonally between bands but UEs experience higher ICI because the Orthogonality between cell edge band of one cell and cell-interior bands of its neighbouring cells is not guaranteed. Strict FFR as shown in Fig.4 is used mostly in multi cell networks and is actually a modification of traditional Frequency reuse with a cell-edge reuse factor. In Fig.5, for the three cells illustrated, the cell centre area is allocated the common sub-band while at the cell-edge remaining bandwidth is portioned with a reuse factor 3. Here, cell centre users don't share spectrum with exterior user and reduce cross tier interference

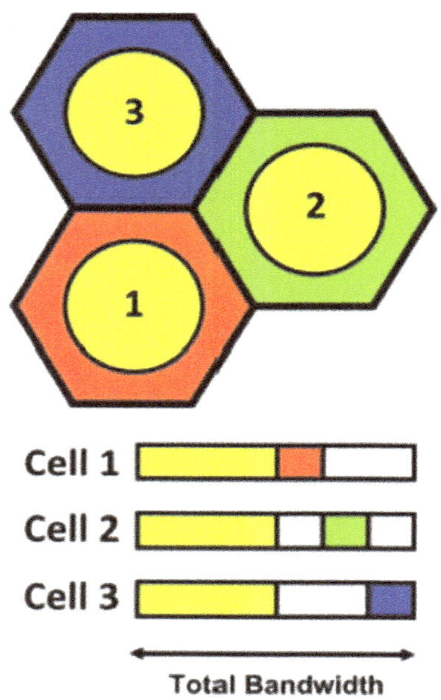

Fig.5: Strict FFR [14]

In reuse factor 3, the whole frequency band can be divided into three equal, orthogonal sub-bands, assigned to sectors so that adjacent sectors always use different frequencies as shown if Fig. 6. FFR3 leads to low improved ICI but large capacity is lost because only one third of the resources are used in each cell.

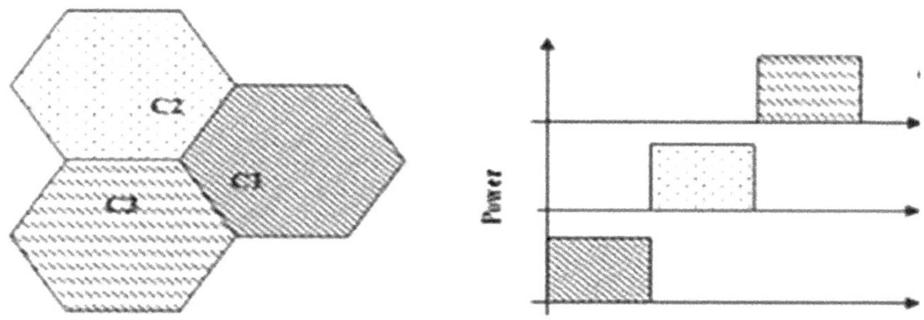

Fig.6: Fractional Frequency Reuse-3

3.LTE-SIM: FEMTOCELL SIMULATION

3.1LTE-Sim Background

LTE-Sim is a LTE network system simulator which written in well-known object-oriented C++ platform [3]. It has the functionalities of simulating several aspects of LTE network, such as E-UTRAN and EPC, downlink and uplink transmissions, single and multi-cell environments, QoS management, multiuser environment, user mobility, handover procedures, and frequency reuse techniques. In addition, the entire LTE protocol stack is framed from the application to the PHY layer, such as the Radio Resource Control (RRC), Radio Link Control (RLC) and MAC entities. Furthermore, *LTE-Sim* supports scheduling strategies (such as Proportional Fair, Log and Exp/PF rules). Moreover, *LTE-Sim* is operating on shell script. After building up the scenario in C++, shell script program works as an interface to start the simulation and analyse the simulation results.

3.2Creating Femtocell Building in simulator scenario

Creating simulation scenario with femtocell includes constructing the macro cells, femtocell building, femtocells, macro cell UEs and femtocell UEs in LTE-Sim. FAPs and eNBs are identified by unique IDs and their positions are defined by Cartesian Coordinate program function [3]. Position tracking, information related to all registered UEs, like the UE identifier, CQI feedbacks, uplink channel quality, and uplink scheduling can be simulated by defining the functionalities debug in the global-config document. In *LTE-Sim*, a building is composed of a number of apartments and in this simulation each one defines the area of a given femtocell. According to the definition in [8], two different types of building are developed and used in *LTE-Sim*: (a) Dual Stripe blocks and (b) 5 × 5 apartment grid. The former consists two building rows composed of 10 apartments each. The later is a building of 25 apartments located over a 5 × 5 grid. This apartment grid can also be modified according to the traffic environment. In this paper, a modified apartment grid (3 × 3) as Fig. 7 is considered for simulation. Each apartment has a squared form and an area of $100m^2$. This means, for instance, a building with 3 × 3 apartment grid can contain up to 9 femtocells. Grid of apartments inside of the building could be defined by configuring the grid settings (i.e. building types) in the header file (SingleCellWithFemto.h). Building walls are formed in NetworkManager.cpp where, function

NetworkManager::CreateBuildingForFemtocells() gets building ID defined in SingleCellWithFemto.h and creates the building according to the building type as presented in Fig. 7.

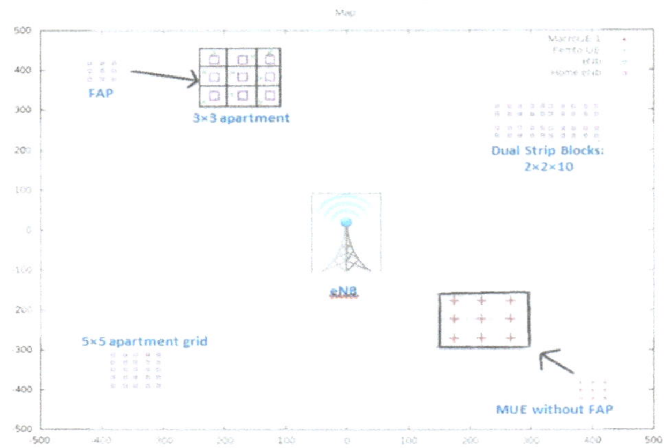

Fig. 7: Femtocell design

```
For(int idBuilding = 0, idBuilding < 1, idBuilding++)
{
    double offsetbuildingCenter_X [1] = {0 };
    double offsetbuildingCenter_Y [1] = {radius/2*sqrt(3)*1000};
    buildingCenter_X = offsetbuildingCenter_X [idBuilding];
    buildingCenter_Y = offsetbuildingCenter_Y [idBuilding];

    nm->CreateBuildingForFemtocells( idBuilding, buildingType, apartmentSide, nbFloors,
                                     buildingCenter_X, buildingCenter_Y,
                                     firstFemtoinBuildingID, femtoCellsInBuilding);

      std::cout << "Created Building, id " << idBuilding
                <<", position: " << buildingCenter_X
                << " " << buildingCenter_Y
                << " and " << nbFloors << " floors"
                << " and " << femtoCellsInBuilding << " femtocells" << std::endl;
}
```

Code Block.1: Positioning of Building containing Femtocells

Code Block.1 shows the positioning of a building containing femtocells (here, the building location is set at X=0, Y=866 coordinates) where, the coverage area of eNB is considered 1000m.

3.3Create UE in simulator

```
double speedDirection = 1.57; //                90/360*2*pi = 1.57
UserEquipment* ue = new UserEquipment (idUE,
                posX, posY, speed, speedDirection,
                cells->at (j),
                eNBs->at (j),
                Handover_index, //HO deactivated if HO_index =0!
                Mobility::RANDOM_DIRECTION);

                std::cout << "Created UE - id " << idUE << " position " << posX << " " << posY
                  << " cell " << ue->GetCell ()->GetIdCell ()
                  << ", target enb " << ue->GetTargetNode ()->GetIDNetworkNode () << std::endl;
```

Code Block.2: Mobility of UE in different HO scenarios

In Code Block 2, it is shown how to create an user at coordinte (0,0) with mobility to perform different HO in different scenarios. In our simulation we considered Random Direction[12] as our mobility model. Here, the UE is moving North, as shown in Fig.8 will pass through femtocell building and handover form Macrocell 1 -> Femtocell 1 -> Femtocell 2 -> Femtocell No.3 -> Macrocell 2.

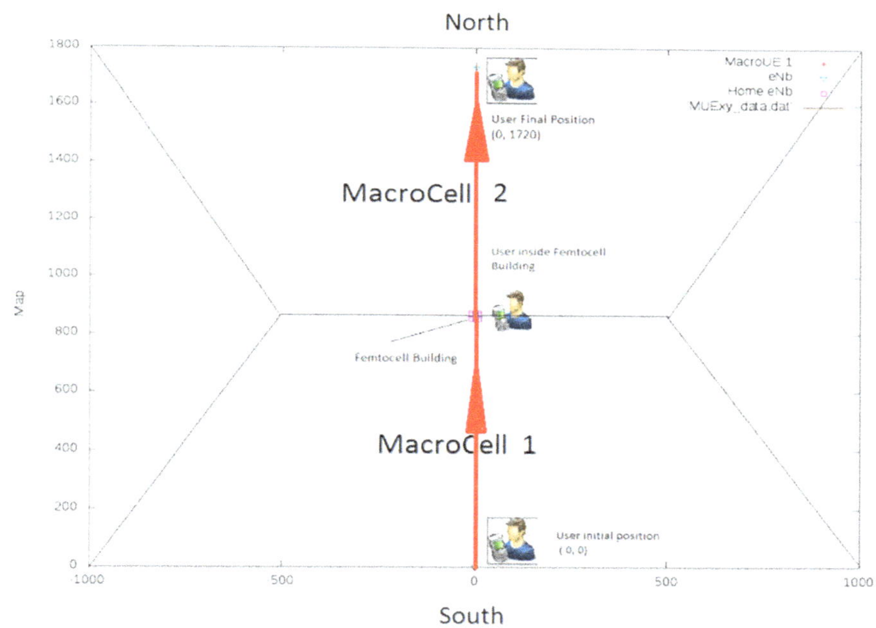

Fig.8: Mobility Pattern of the UE

4.SIMULATION RESULT

In this section, some simulation outcomes regarding the change in SINR of UE in different coverage area with and without presence of Femtocells are shown. The simulation results show: (i)SINR improvement of the UE as it moves from macro-cellular coverage to femto-cellular coverage area when FFR was not implemented, (ii) the SINR of the user further improves when FFR was implemented (iii) a scalability test on computational requirements (i.e., simulation time and CPU usage) was also performed at the end. Major simulation parameters are summarized in Tab. 1. In our simulation, presences of two eNBs were considered along with the UE having the access permission to all the femtocells. Building containing the femtocells in a 3 × 3 pattern is located at the cell edge of two cells.

Table 1: Simulation Parameters

Total Bandwidth	5 MHz
eNb power transmission	43 dBm
FAP power transmission	20 dBm
CQI	Full Bandwidth and periodic (2ms) reporting scheme
Apartment Size	100 m^2
Mobility Model	Random Direction
HO Type	Position Based HO
Femto cell/Apartment	1/1
Frequency Reuse	Soft FFR, Strict FFR, FFR3
Apartment orientation	3 × 3
Buildings	1
Users	1
Scheduler	PF
Traffic	VOIP, VIDEO

As we can see from the simulation result, illustrated in Fig. 9 that, when the UE is moving from its initial position of Macrocell 1 coverage area to cell edge both the uplink and downlink SINR is decreasing because of path-loss, penetration and macro-macro co-channel interference. Position based HO model is considered in this simulation. As the UE moves closer to the FAP, HO takes place from eNB to FAP and it can be seen from Fig. 9 that the SINR has rapid improvement inside the FAP coverage area. Since we considered 3 × 3 apartment grid, it is also seen from the figure that, UE passed three FAP coverage area and 3 successful HO attempts were made. The fluctuation (three peaks in SINR graph inside FAP area) in SINR seen if Fig. 9 is because of the handovers to three different femtocells. As the UE passed the FAP coverage area, SINR has upward trend since, UE is moving towards the cell centre area of the Macrocell 2 coverage area.

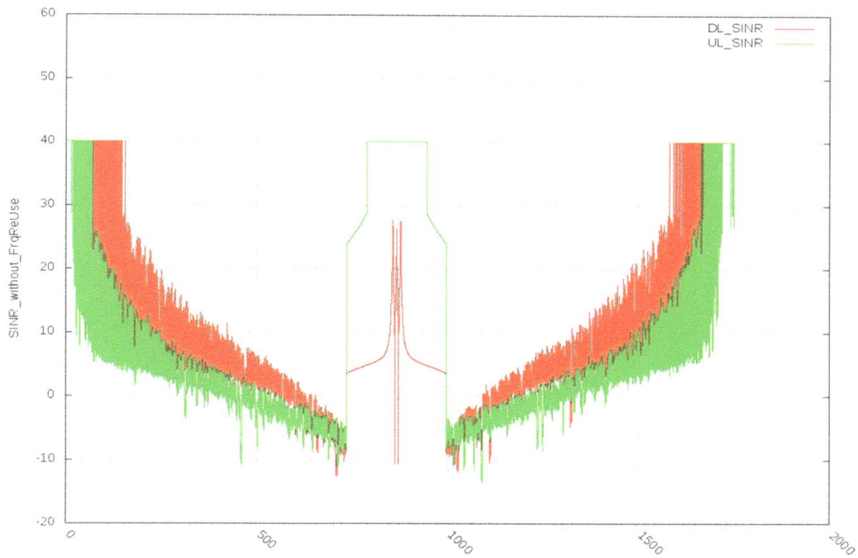

Fig.9: SINR without Frequency Reuse

When FFR is implemented DL SINR improves, which can be seen in Fig.10. FFR has no effect on uplink SINR since single UE is considered with no interference just path –loss, penetration-loss, multipath and shadowing.

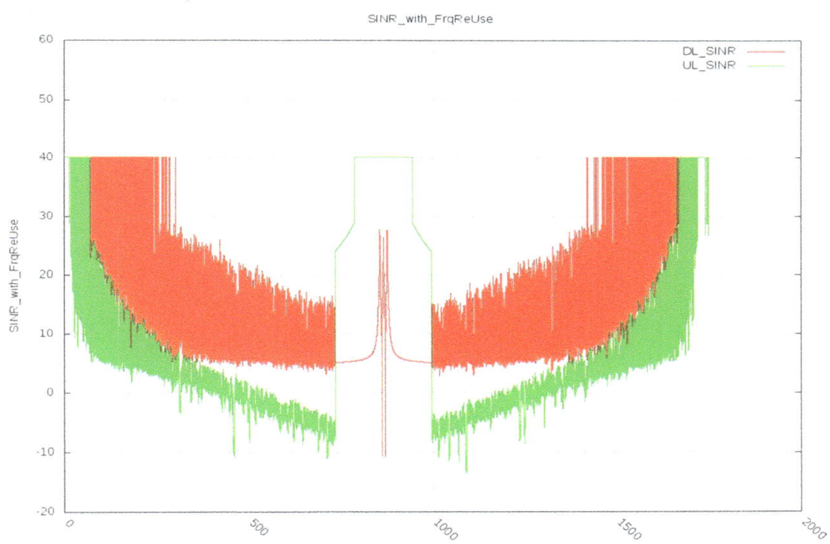

Fig. 10: SINR with Frequency Reuse

For more performance evaluation in different frequency reuse schemes i.e., FFR-3, Soft Frequency Reuse and Strict Frequency Reuse in with presence of increasing number of femtocells (maximum of 40 femtocells) are installed randomly inside the of the macrocell coverage. Fig. 11 shows the outage probability in different reuse schemes in different threshold SINR. Strict FFR has better performance than the other two this is because in Strict FFR the cells centre FUEs are not interfered by the FUEs of the edge area.

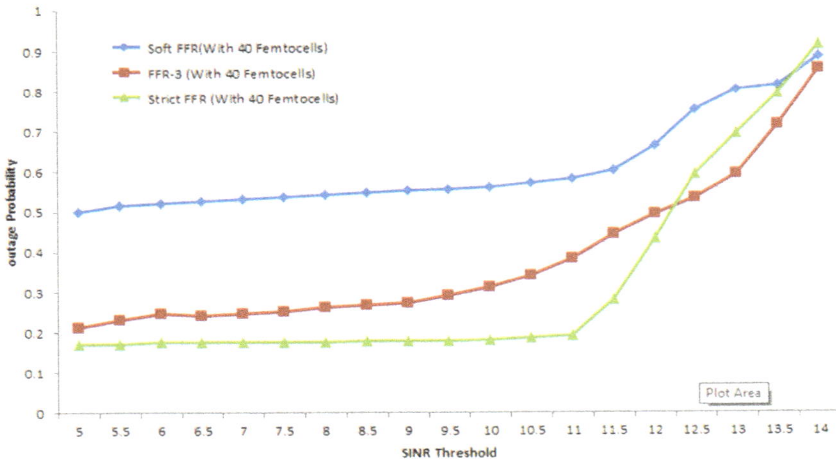

Fig. 11: Outage Probability Vs SINR Threshold in presence of Femtocells

The change in network throughput with the increase in number of femto cells is shown in Fig.12. With the increase in femtocell number throughout the macrocell coverage the network throughput decreases because of co-tier and cross tier interferences. FFR-3 schemes show the better network throughput.

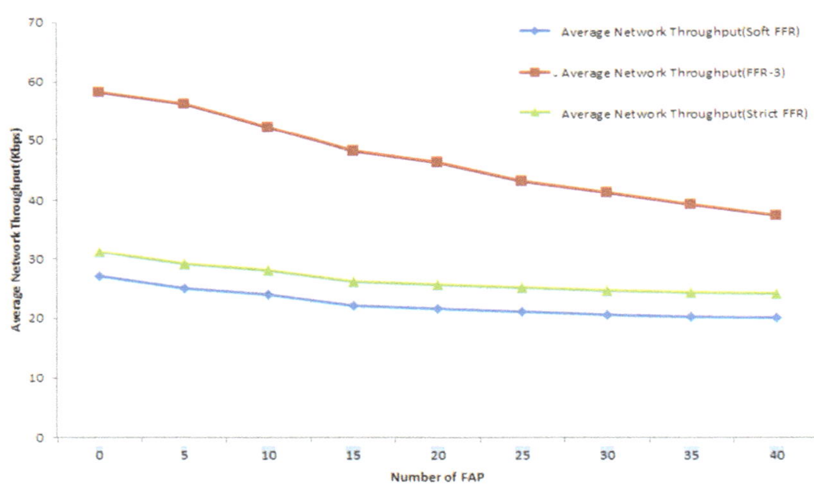

Fig. 12: Variation of Network Throughput with femtocell numbers

Fig.13 shows the spectrum efficiency of different reuse schemes in both cell centre(along with edge) and edge. FFR-3 has the better spectral efficieny in both the case.

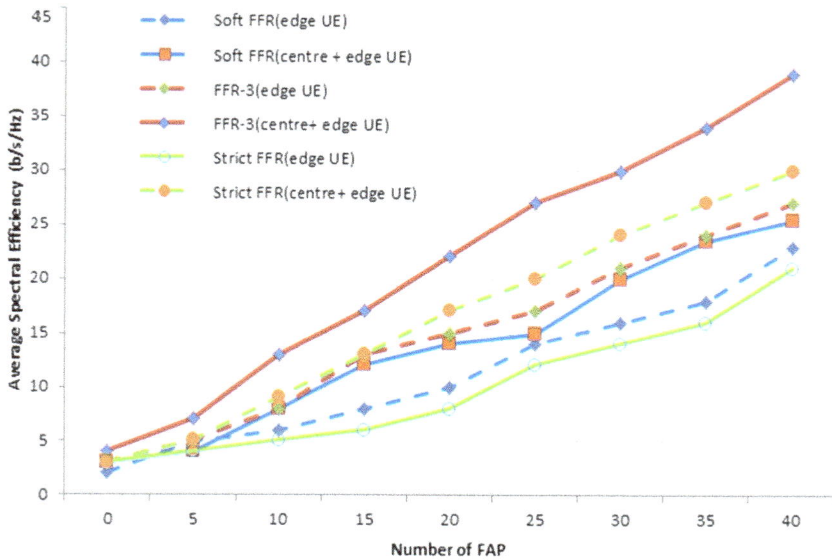

Fig. 13: Spectral Efficiency of different frequency Reuse schemes

The scalability test of the presented simulation model was running on Linux machine with a 3.2 GHz (Dual Core) CPU and 8Gbytes of RAM. Obtained measurements in Fig. 14 shows the simulation time and memory usage linearly increases with the increasing number of users and it can be seen in Fig. 14c that the CPU usage (1 out of 4 cores was fully used) remains the same in the scenarios with the highest computational load, which proofs the scalability of the presented model in *LTE-Sim*.

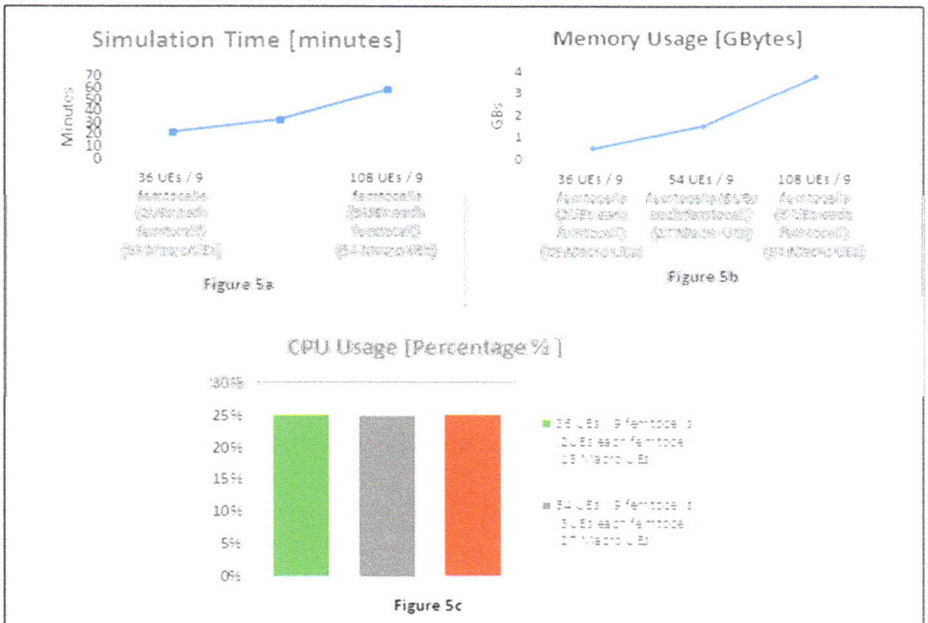

Figure 14: Simulator Performance in terms of a) Simulation Time b) Memory Usage c) CPU Usage(percentage %)

5. CONCLUSION

In this article, demonstration to simulate femtocell scenarios with features and models compliant with 3GPP specifications in LTE-Sim framework is provided. The femtocell scenario shown in this paper can be used for investigating performances in a wider range that can differ in terms of network layouts (i.e., number and position of base stations, building, and streets), system load (i.e., number of users and applications) and so on. As a future study, we plan to enhance the *LTE-Sim* Femtocell framework with the aim of implementing mobility management, Handover, Radio Resource Allocation schemes [13] with random users both in eNB and FAP, handover, radio resource allocation schemes [21] as well as other algorithms and protocols based on both self-organizing network (SON) and cognitive paradigm believing the extreme modularity of *LTE-Sim*.

REFERENCES

[1] L. Nuaymi, WiMAX: Technology for Broadband Wireless Access. Wiley, NewYork, 2008.

[2] E Dahlman, S Parkvall, J Skold, P Beming, 3G Evolution HSPA and LTE for Mobile Broadband. Academia Press, USA, 2008.

[3] F. Capozzi, G. Piro, L. Grieco, G. Boggia, P. Camarda, "On Accurate simulations of LTE femtocells using an open source simulator," in EURASIP Journal on Wireless Communication and Networking, 2012.

[4] G Piro, L Grieco, G Boggia, F Capozzi, P Camarda,"Simulating LTE cellular systems: an open-source framework, " in Vehicular Technology, IEEE Trans. vol. 60, pp. 498-513, 2011.

[5] Raheem, Rand, Aboubaker Lasebae, and Jonathan Loo. "Performance Evaluation of LTE Network via Using Fixed/Mobile Femtocells." Advanced Information Networking and Applications Workshops (WAINA), 2014 28th International Conference on. IEEE, 2014.

[6] WCDMA for UMTS – HSPA Evolution and LTE, Fourth Edition. Edited by Harri Holma and Antti Toskala © 2007 John Wiley & Sons, Ltd. ISBN: 978-0-470-31933-8

[7] D. Lopez-Perez, I. Guvenc, G. de la Roche, M. Kountouris, T. Q. S. Quek, and Z. Jie, "Enhanced intercell interference coordination challenges in heterogeneous networks," IEEE Wireless Commun., vol. 18, pp. 22-30, 2011.

[8] D. Xenakis, N. Passas, L. Merakos, C. Verikoukis, "Mobility Management for Femtocells in LTE-Advanced: Key Aspects and Survey of Handover Decision Algorithms" in IEEE Trans. Wireless Commun., vol. 18 pp. 2011.

[9] R. Bendlin, V. Chandrasekhar, C. Runhua, A. Ekpenyong, and E. Onggosanusi, "From homogeneous to heterogeneous networks: A 3GPP Long Term Evolution rel. 8/9 case study," in 45th Annu. Conf., Inform. Sci. and Syst. (CISS), Baltimore, MD, pp. 1-5, 2011.

[10] H. Mahmoud, I .Guvenc," A comparative study of different deployment modes for femtocell networks," in IEEE Int. Symposium on Personal, Indoor and Mobile Radio Commun. , Palo Alto, USA, pp. 1–5, 2009.

[11] Siemens, R1-060135, "Interference Mitigation by Partial Frequency Reuse," 3GPP RAN WG1#42, London, UK, January 2006.

[12] T. Camp, J. Boleng, and V. Davies, "A survey of mobility models for ad hoc network research," Wireless Communications and Mobile Computing, vol. 2, pp. 483–502, Aug. 2002.

[13] F. Capozzi, G. Piro, L. Grieco, G. Boggia, P. Camarda, " A system-level simulation framework for LTE Femtocell," in Proc. of International Conference on Simulation Tools and Techniques(ICST), Desenzano, Italy, pp.1-3, 2012.

[14] http://www.infoutils.com/frequency-reuse-in-ofdma-based-ltewimax-system/

AUTO DEFAULT GATEWAY SETTINGS FOR VIRTUAL MACHINES IN SERVERS USING DEFAULT GATEWAY WEIGHT SETTINGS PROTOCOL (DGW)

Suman Dutta[1], Shouman Barua[2] and Jishu Sen[3]

[1]IT Trainer, Logitrain.com.au
[2] PhD research student. University of Technology, Sydney
[3] MSc research student. University of Technology, Sydney.

ABSTRACT

Availability is one of the most important concerns in the networking world. For any high available network, we need to maintain 99.99999% availability. That is why it is one of the most important factors to find out the single point of failure in the network architecture and eliminate that single point of failure (SPOF) from physical network and logical network. SPOF in our server infrastructure has been analysed in terms of communicating with the router for forwarding traffic with multiple routers. We have developed an algorithm that will automatically select default gateway into the network interface card of virtual machines. The proposed algorithm will automatically enable Default Gateway Weight settings (DGW) protocol among routers by configuring Network interface card with default gateway of all routers. The proposed protocol works based on weight settings for the multiple default gateway configuration in the host. There will be heartbeat communication and re-convergence will be performed within the shortest possible time. Lowest weight setting will select the path for packet forwarding through specified routers related with the default gateway from the virtual machine.

KEYWORDS

DGW protocol, Auto default gateway, Virtual Machine, High Availability, Weight settings.

1. INTRODUCTION

Virtual machine is the integral part to work as a server for different applications. In order to provide high availability for the virtual machine traffic, we can use network interface card (NIC) teaming switch independent mode for virtual network adapter of virtual machines. Virtual network interface card of virtual machines will work with automated default gateway settings using weight configuration among the virtual network adapter which will provide high availability among the virtual machines for application that will run into virtual machine. We will use an algorithm that will run with the protocol to converge the network with virtual machines for achieving high resiliency with multiple router and will resolve the problem of single point of failure. To implement our high reliability with virtual machines through automatic default gateway settings, we will use many networking and server virtualization concepts.

2. CONCEPT OF VIRTUALIZATION

Through the last decade, the enterprise data centre infrastructure design has been changed in a large scale to provide multifarious services. It is now an integral part of data centres to provide services to numerous clients including roaming devices, who are not only connected to the data centre from internal network of organization, but also connected from other geographic locations outside the organization. It means that organizations now need to provide much higher levels of availability and redundancy for internal services which will be accessed from the internet. Successful organizations generally can adapt rapidly to change business requirements, and can attain the high availability and resiliency.

2.1. High Availability with virtualization

The enterprise data centre provides services to different of potential clients, including organization employees who are internal and external, customers and business partners. From any place and at any time, the clients may need to access the data center service. Because of this, the most pivotal task is to make the service highly available and responsive. In the case of a service outage due to the network failure, it is very critical that the service can be restored within a shortest possible time frame.

To provide a highly availability, it is noteworthy that the system can failover or recover from the loss of any single component of the network. The system should be able to detect whenever a component will fail and there should be a mechanism to automatically restore from that failure. Generally, this is achieved by shifting the services from the failed component to a redundant component. It is also important that this failover should be happened automatically and that it does not require administrative attention or effort. But running Virtual machines for providing different services using multifarious applications subject to the constraint of high availability.

2.2. Virtual Switch

A virtual switch is a software program that works like a physical Ethernet switch that allows multiple virtual machines (VM) to inter communicate with each other. There is a difference between physical Ethernet switch and a virtual switch. Virtual switch is not only forward data packets, but it can intelligently direct communication on the network by inspecting packets before passing through the network. Few vendors has embedded virtual switches right into the virtualization software, but a virtual switch can also be included in a server's hardware as a firmware. Because of its intelligence, virtual switch can ensure the integrity of a VM profile that includes network and security settings.

2.3. NIC teaming

In NIC teaming solution, multiple physical network adapters are connected to the NIC Teaming solution-multiplexing unit. This multiplexing unit presents one or more virtual adapters, which is also known as team network adapters to the operating system. There are different algorithms, which distribute inbound and outbound traffic between the physical network adapters. In non-Microsoft NIC Teaming solutions allows the team network adapters to divide traffic based on virtual local area network (VLAN) so that applications can connect to different VLANs.

2.4. Default Gateway

In computer networking, a default **gateway** is a router on a network, which serves as an access point to another network through which computers from a LAN network can communicate with outside world. In home computing network configurations, an ISP provides a physical device that both connects local network to the Internet and serves as a gateway. Such devices are digital subscriber line (DSL) routers and cable routers.

In organizational network infrastructure, a gateway is a node, which is used to routes the data packet from a workstation of one network to another network. The default gateway generally connects the internal networks and the outside network i.e., internet.

2.5. Requirements for Redundant Default Gateway Settings

When a host needs to communicate with a device on its local subnet, it needs to generate a broadcast message throughout the subnet which is an Address Resolution Protocol (ARP) request and wait for the unicast address resolution protocol reply and exchange packets. However, if the destination device is located on a different subnet then the sending device depends on an intermediate system (a router, for example) to forward packets to and from that subnet.

A sending workstation identifies its nearest router, which is also known as the default gateway or next hop, by its IP address. If the sending device comprehends something about routing, it considers that all packets destined off-net must be sent to the gateway's media access control (MAC) address instead of the other end's MAC address. Therefore, the sending device first sends an address resolution protocol (ARP) request to find the default gateway's MAC address. Then packets can be sent to the default gateway directly without looking for ARP entries for individual destinations which has been illustrated in Figure 1.

If the default gateway router becomes down, then hosts cannot communicate with the networks that is connected with the router. To avoid single point of failure, we can implement another default gateway within the same subnet to eliminate single point of failure that has been illustrated in Figure 2.

But there is a problem with this type of settings. If one router goes down, then we have to manually change the default gateway in the client PC's to forward packet through another default gateway router within the same LAN. To avoid this problem, we use our algorithm with new protocol default gateway weight settings (DGW).

2.6. **Existing protocol that are used for redundant default gateway:**

Hot standby router protocol is one of the protocol (HSRP) that is used to operate with an active/standby model. Two routers cooperate with each other using HSRP protocol to act as a default router. However, only one protocol will work as a default gateway at any time to actively support the end user traffic. The active router will carry the end user traffic for the other network. Then the other routers of the HSRP protocol will be in stand by state. It will wait to take over the control from the active router if active router face any problem which has been showed in Figure 3.

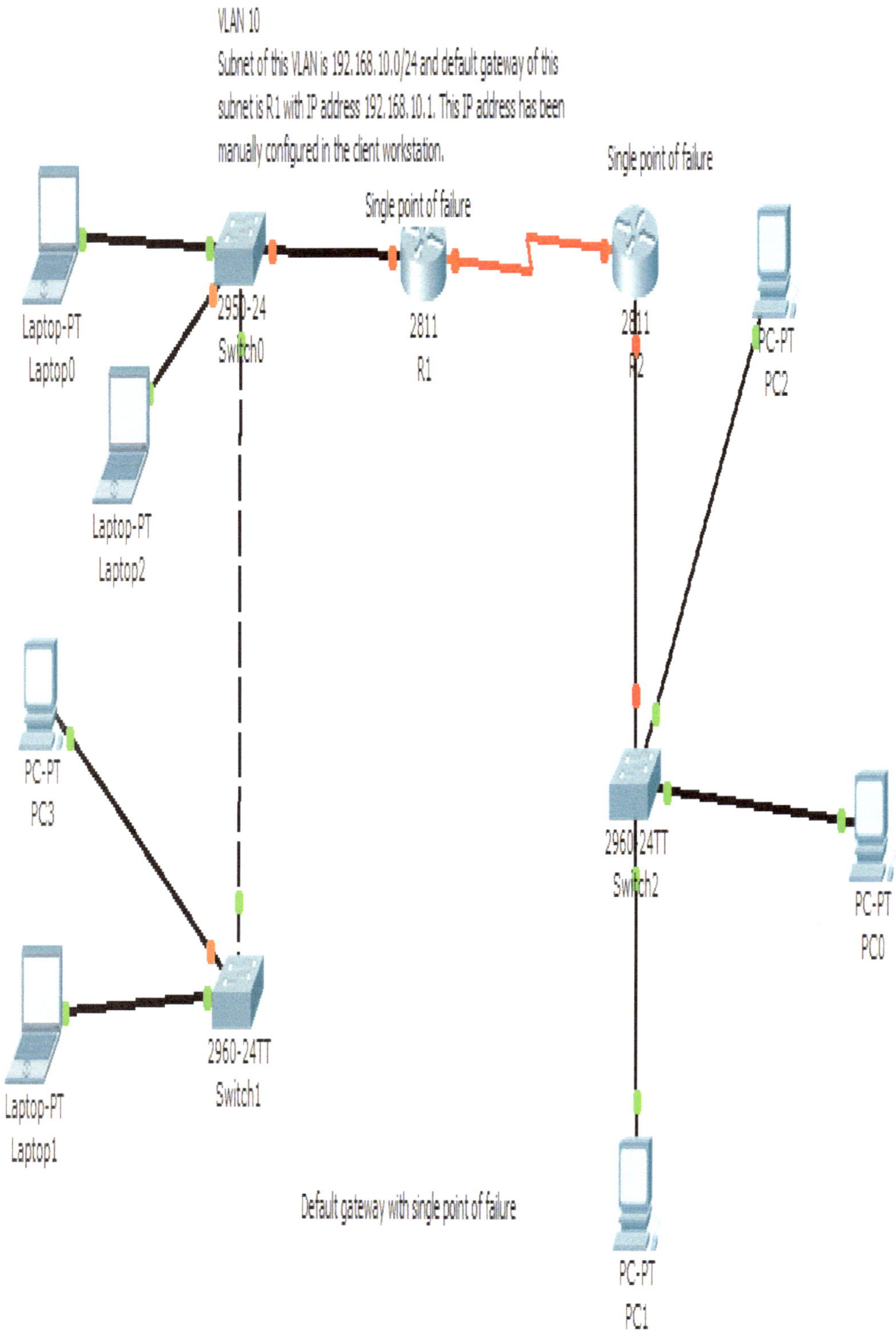

Figure 1: Default gateway with single point of failure

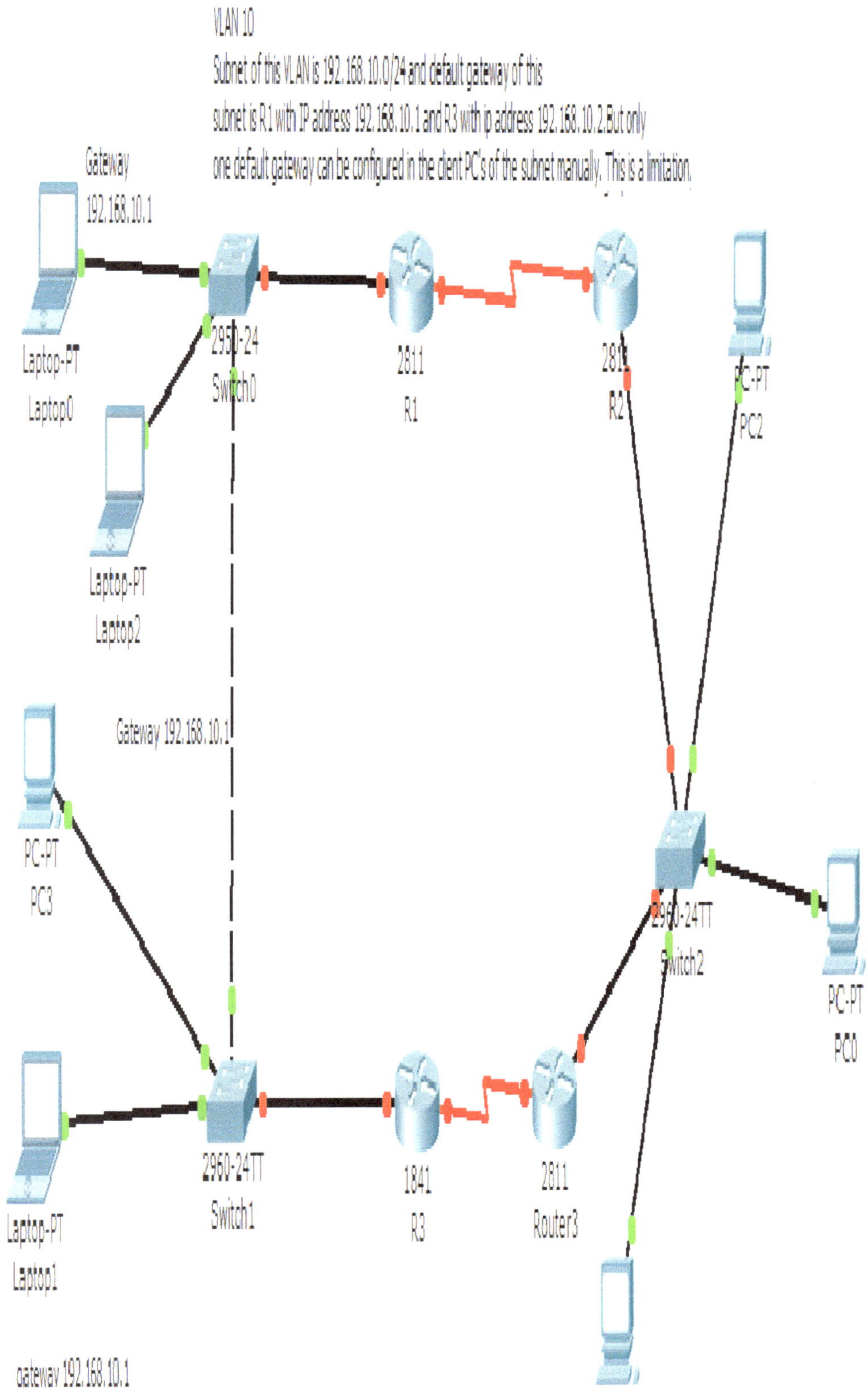

Figure 2: Redundant default gateway with manual default configuration

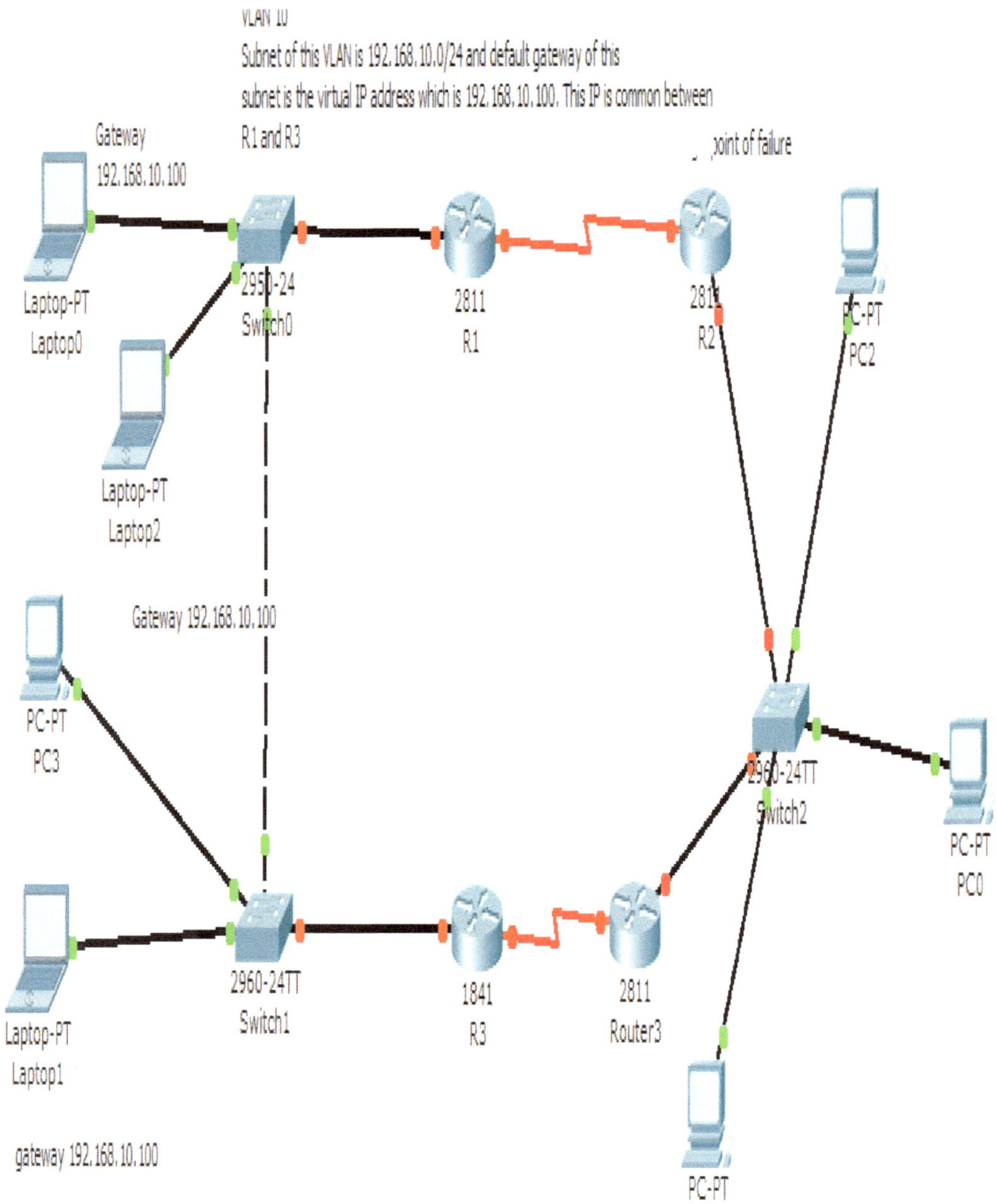

Figure 3: Default gateway with HSRP protocol

The HSRP protocol works with the virtual IP address. Active router controls the virtual IP address. This virtual IP address is part of the HSRP configuration. This IP address is the additional address in the fast Ethernet interface of the active and standby routers. All the IP address belongs to the same subnet. The router then automatically creates virtual MAC address. All the co-operating HSRP routers know these virtual address, but only active default gateway will use virtual IP address at any one point of time.

2.7. DGW protocol

DGW protocol needs to be enabled into the routers Fast Ethernet interface that is connected with the local area network (LAN) switch. The number of routers that would be connected with the LAN switch should be minimum two. Host Server will be connected with switch thorough multiple network interface cards. All network interface cards should be combined together to form virtual network adapter, which is also known as the NIC teaming to provide the fault tolerance and high availability for the server.

The server is running with hypervisor where multiple virtual machine can be configured for different service. To implement our protocol for automatic default gateway, all IP address should be configured within a single subnet of all virtual machines and routers Ethernet interface of the LAN which has been showed in Figure 4.

On the network interface card of virtual machine, we need to configure the options for our DGW protocol for Ethernet interface that needs to be performed with the driver software update of different vendor specified network interface card.

When we enable our protocol into the router Ethernet interface and virtual machine Ethernet interface, this will generate a multicast hello message within that subnet using the default multicast address 224.254.254.254. With this hello messages routers will send their LAN interface IP address configuration. Every router will analyze the hello messages to identify whether all IP address of the routers belong to the same subnet in the LAN network. Using the multicast message, DGW protocol will work between the routers. DGW protocol will select only one active router and standby routers based on the uptime of routers. Uptime will be calculated from the routers. The routers with the highest uptime will become the active router and weight will be set to 0 and for the standby routers weight will be set according to the chronological orders. For example 1, 2, 3…etc. Now this auto weight settings of the routers will be send to the virtual machines though the multicast message of neighbor advertisement using DGW protocol. Here the multicast message would be 224.253.253.253.

DGW protocol will always run on the background of network interface card (NIC) card of the virtual network adapter of the virtual machines. The neighbor advertisement message of the DGW protocol will put the default gateway settings. In the NIC card, there should be the options of multiple default gateway settings with different weight i.e., IP address with weight

The DGW protocol, which will also run in the network interface card, will be used to route packets with the lowest weight settings of the router IP address. DGW protocol will always send multi casted keep alive messages using the multicast address 224.253.253.253 to active default gateways at every 100 millisecond. If two keep alive messages are missed between host virtual machine and active default gateway router, then DGW protocol will automatically choose next lowest weight settings as the default gateway in the virtual NIC card of virtual machine. At the same time, this router with the weight settings 1 will become the active router and all other router will become the standby routers.

In this way, virtual machine will provide the fault tolerance with automated default gateway setting using DGW protocol with maximum down time of 200 milliseconds.

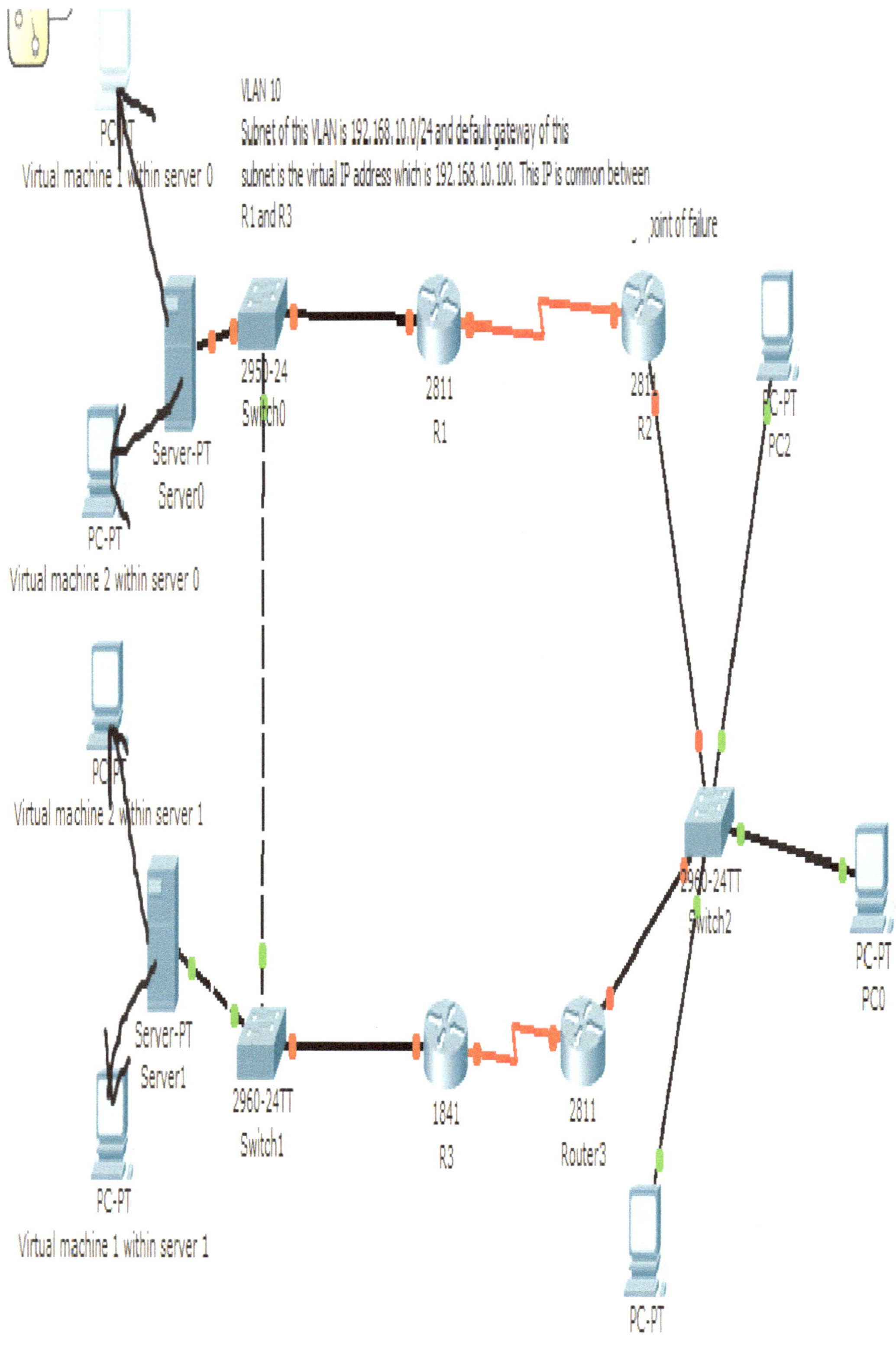

Figure 4: DGW protocol with redundant default gateway.

same time, this router with the weight settings 1 will become the active router and all other router will become the standby routers.

In this way, virtual machine will provide the fault tolerance with automated default gateway setting using DGW protocol with maximum down time of 200 milliseconds

2.7. Pseudo code of DGW:

Router_weight_settings()

{

1. The routers will generate multicast Hello messages using IP address 224.254.254.254 using the Ethernet interface.
2. Every router will analyze their network address by calculating through the subnet mask. If all routers will belong to the same network address then they will establish neighbor ship with each other.
3. Collect the uptime of each router by using the multicast messages.
4. Everybody will compare their own uptime with the uptime of the other routers and the higher one will win.
5. Assign weight 0 with the IP address of the highest uptime of router.
6. Increment the weight number or next higher uptime routers and continue it for all remaining routers.
7. Router with lowest weight settings will hold a copy of weight settings of all routers.
8. Send output to the function weight_settings

End of function.

}

Weight_settings()

{

1. Contains the list of outputs.

}

Multicast_message_generation_virtual_machine()

{

1. Virtual machines in server generates muiticast message with destination IP 255.253.253.253.
2. Only routers will sense that multicast message and determine whether source IP address belong to the same network of the routers LAN interface.
3. If source IP address belong to the same network, then
4. {
5. Lowest weight settings router will send a unicast reply messages to the virtual machine of the same network which will contain list of default gateways with different weight settings virtual machine function

   }
6. All virtual machines within the same network will get the list of default gateways.
7. Selecting default gateway by invoking the function select_default()

}

Select_default()

{

1. Select the lowest weight settings of the default gateway list and place lowest weight settings as the default gateway of that virtual machine network adapter.
2. Start function monitoring_default_gateway()

}

Monitoring_default_gateway()

{

1. Start sending and receiving keep_alive_messages between host and selected default gatewayat every 100 millisecond
2. If two consecutive keep_alive_messages are missed then
3. Track the default gateway with the lowest weight settings.
4. Send this default gateway to the function calculate_new_gateway()

}

Calculate_new_gateway()

{

1. Invoke function weight_settings().
2. Select next lower default_gateway from the list of default_gateway.
3. Invoke function Monitoring_default_gateway()

}

2.7. Comparison between DGW protocol and other existing protocols

There are copious advantages of using this DGW protocol over the current hot standby redundant protocol (HSRP) and Virtual Redundant router Protocol (VRRP). Both HSRP and VRRP protocol supports active and standby mode of routers. These protocol requires virtual IP address and needs manual configuration into the routers. Moreover, these protocol can be implemented only with two routers. Furthermore, to switch between the routers, it takes 3-5 seconds. The DGW protocol will automatically start functioning with the routers and hosts. There is no requirement of manual default gateway configuration in the hosts. This protocol is not limited to the number of routers and it can be re-converged within 200 milliseconds. So, DGW protocol provides seamless redundancy and fault tolerance among the virtual machines of the servers which ensures high availability of applications and service through virtualization. In spite of being planned for virtual machines, this protocol can also supports physical machines.

3. Challenges For Implementing DGW Protocol

Network interface card driver need to reprogrammed for supporting the algorithm of this protocol and router interworking operating system (IOS) also needs to be updated to support this protocol.

4. Conclusions

The main purpose of the default gateway weight setting protocol is to ensure the high reliability, availability, and high fault tolerance for the default gateway of virtual network adapter settings in the virtual machines of the host. The DGW protocol will work at the virtual network interface card of virtual machine as well as physical network interface card of physical machines.

REFERENCES

[1] Cisco. Nexus 1000V Series Switches. http:// www.cisco.com/en/US/products/ps9902,July 2009.

[2] P. Congdon. Virtual Ethernet Port Aggregator Standards Body Discussion. http://www.ieee802.org/1/files/ public/docs2008/new-congdon-vepa-1108-v01. Pdf, Nov.2008.

[3] P. Barham, B. Dragovic, K. Fraser, S. Hand, T. Harris, A. Ho, R. Neugebauer, I. Pratt,and A. Warfield. Xen and the Art of Virtualization. In Proc. of SOSP, Oct. 2003.

[4] A. Bavier, N. Feamster, M. Huang, L. Peterson, and J. Rexford. In VINI Veritas: Realistic And Controlled Network Experimentation. In Proc. of SIGCOMM, 2006.

[5] M. Casado, T. Koponen, D. Moon, and S. Shenker. Rethinking Packet Forwarding Hardware. In Proc. of HotNets, Nov. 2008.

[6] Altar Networks. http://altornetworks.com, July 2009.

[7] T. Anderson, L. Peterson, S. Shenker, and J. Turner. Overcoming the Internet Impasse through Virtualization. Computer, 38(4):34–41, 2005.

[8] F. Anhalt and P. V.-B. Primet. Analysis and Evaluation of a Xen Based Virtual Router. Technical Report 6658, Inria, Sept. 2008.

[9] R. Davoli. VDE: Virtual Distributed Ethernet. In Proc. Of TRIDENTCOM, Feb. 2005.

7

ANALYSES AND PERFORMANCE OF TECHNIQUES PAPR REDUCTION FOR STBC MIMO-OFDM SYSTEM IN (4G) WIRELESS COMMUNICATION

Leila Sahraoui, Djmail Messadeg, Nouredinne Doghmane

Department of Electronics Faculty sciences of Engineering
University Baji Mokhtar, Annaba bp 12 el hadjar, Algeria

ABSTRACT

An OFDM system is combined with multiple-input multiple-output (MIMO) in order to increase the diversity gain and system capacity over the time variant frequency-selective channels. However, a major drawback of MIMO-OFDM system is that the transmitted signals on different antennas might exhibit high peak-to-average power ratio (PAPR).In this paper, we present a PAPR analysis reduction of space-time-block-coded (STBC) MIMO-OFDM system for 4G wireless networks. Several techniques have been used to reduce the PAPR of the (STBC) MIMOOFDM system: clipping and filtering, partial transmit sequence (PTS) and selected mapping (SLM). Simulation results show that clipping and filtering provides a better PAPR reduction than the others methods and only SLM technique conserve the PAPR reduction in reception part of signal.

KEYWORDS:

MIMO-OFDM; peak-to-average power ratios; space-time coding system (STBC); clipping and filtering; SLM; PTS.

1. INTRODUCTION:

Orthogonal Frequency Division Multiplexing (OFDM) is multi-carrier modulation (MCM) scheme technique used for 4th Generation (4G) wireless communication. This technique with high-speed data transmission is used in mobile communication, Digital terrestrial mobile communication, Digital Audio Broadcasting (DAB) and Digital Video Broadcasting terrestrial (DVB-T). OFDM has many advantages such as High spectral efficiency, immunity to inter-symbol interference and capability of handling very strong multipath fading [1]. Despite its competitive position attributes, OFDM signals are characterized by very high levels of Peak to average power (PAPR Ratio). This peculiarity conduit OFDM signals to be very sensitive to non-linearities of the analog components of the transceiver, in particular those of the high power amplifier (HPA) at transmission. A HPA is designed to operate in its saturation region which corresponds to a high yield of the region.
However, in this area, the HPA nonlinear behavior severe. These nonlinearities are sources of in-band (IB) distortions that can both reduce the performance in terms of Bit Error Rate link (BER)

[1]. For all these reasons, reducing the PAPR of OFDM signals is increasingly being considered to be very important to preserve the cost-effectiveness advantages of OFDM in practical systems. MIMO system is the use of multiple antennas at both the transmitter and receiver to enhance communication performance. OFDM combined with MIMO technology with the block of space-time coding (STBC) represents an interesting candidate for mobile communication systems due to its ability to withstand high speeds, high capacity and robustness to multipath fading [2]. However, as with single-input single-output OFDM (SISOOFDM), one of the major drawbacks of STBC MIMO-OFDM is the high peak-to-average power ratio (PAPR) of the signals transmitted on different antennas. A straightforward way to reduce the PAPR in STBC MIMOOFDM systems would be to apply the PAPR reduction approach proposed for SISO-OFDM systems [3] to each antenna separately and the side information (SI) required to recover the signal successfully at the receiver would increase proportionally with the number of antennas [4], [5]. A space-time coding system (STBC) with MIMO-OFDM is deployed for transmit diversity and secure means for data propagation in the scenario where the mobility is required for the data transmission. Is less a high PAPR always accompanies MIMO-OFDM systems. To remedy this problem, there are now a multitude of techniques for reducing PAPR have been proven in many applications of digital communications such as wireless networking son (Wi-Fi), the (WiMAX), HiperLAN and recently in the DVBT2. Many methods have been proposed to mitigate the OFDM PAPR by acting on the signal itself [2], [6]. The simplest ones use clipping and filtering techniques [4], [5]. However, these methods may increase the BER of the system since clipping is a nonlinear process [7].

Alternative methods are based on coding [8] and others on Multiple Signal Representation (MSR) techniques: Partial Transmit Sequence (PTS) [9], Selective Mapping (SLM) [10]. The main drawback of these methods is that a Side Information (SI) has to be transmitted from the transmitter to the receiver to recover the original data, which results in some loss of throughput efficiency [7].

Among these techniques, we propose to use a combination of STBC coding techniques (Alamonti) in MIMO-OFDM, with reduction techniques to better discriminate the PAPR reduction induced by the most widely used methods are clipping and filtering, selective mapping (SLM) and partial transmit sequences (PTS).

Our study is based on the PAPR reduction in transmission. Two approaches were made: Reduce the PAPR issue and compare the shape of the complementary cumulative distribution function (CCDF) curve of the PAPR for transmission and reception for each of the three methods mentioned above.

2. STBC MIMO-OFDM system:

We can write the complex vector of size N

$$X_{K} = X_0, X_1, \ldots\ldots\ldots X_{N-1}$$

Where, X_k is the complex value carried by the kth subcarrier. The OFDM symbol can be written as:

$$x(t) = \frac{1}{\sqrt{N}} \sum_{k=0}^{N-1} X_k e^{j2\pi K f_0 t}, 0 \le t \le T \qquad (1)$$

Where T is the symbol interval, and f_0 =1/T is the frequency spacing between adjacent subcarriers.

Replacing t=n T_b, where T_b =T/N, gives the discrete time version denoted by

$$x(n) = \frac{1}{\sqrt{N}} \sum_{k=0}^{N-1} X_k e^{j2\pi kn/LN}, n = 0,1,......NL-1 \qquad (2)$$

Where, L is the oversampling factor.

Fig. 1 illustrates the general block diagram of a STBC MIMO-OFDM system. Baseband modulated symbols are passed through serial-to-parallel (S/P) converter which generates complex vector of size *N*. Then the complex vector, *X* is then passed through the STBC encoder. Both these sequences are then passed through each IFFT block for antenna 1 and antenna 2 respectively.

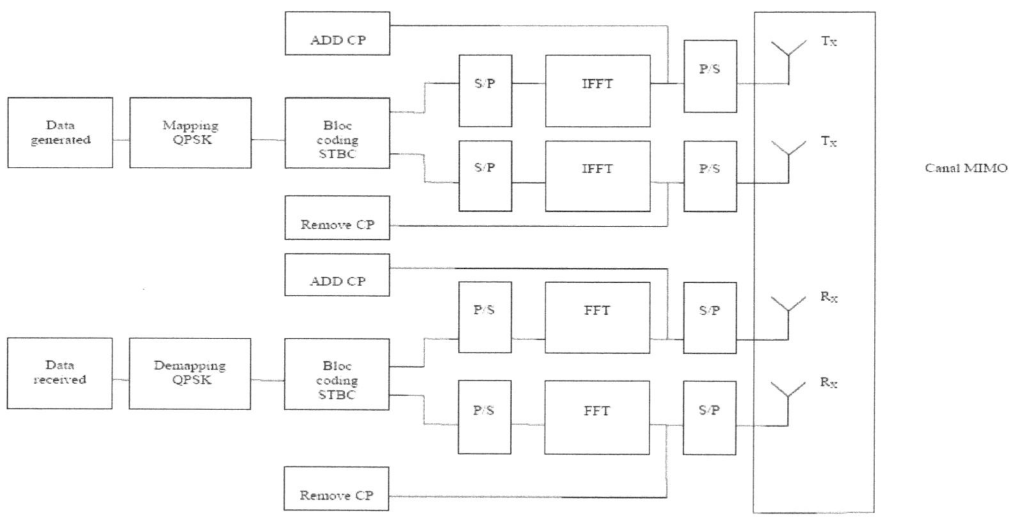

Figure 1. Structure of STBC MIMO-OFDM

2.1. Alamouti Space-Time Block Code (STBC):

Increasing the transmission rate and providing robustness to channel conditions are nowadays two of the main research topics for wireless communications. Therefore, a significant interest of late has been to develop systems that offer both high capacity and high data speed. Indeed, much effort is done in the area of Multiple Input Multiple Output (MIMO) systems by using several antennas either at the transmitting side or at the receiving side. We can exploit space and time diversity by using Space time codes such as the famous Alamouti code In multi antennas communication systems, investigations of *Nt* transmit and *N r* receive antennas showed that the capacity of such systems increases linearly with the minimum of *Nt* and *Nr*. High data rates are obtained by simultaneously sending signals from several transmit antennas [11].

The space-time coding technique is essentially a two dimensional space and time processing method. While multiple antennas both for transmission and reception are used to improve wireless communication systems capacity and data rate in space-domain, in time-domain, different signals can be transmitted at different time slots using the same antenna at the same time. Correlation of time and space is introduced between signals which are transmitted by different antennas so that the receiver antennas can realize diversity reception.

Therefore, space-time coding is especially meant for higher coding gain without using more bandwidth which effectively enhances capacity of wireless systems [12], [13], [14].

In our paper, we have used the Alamouti space time coding to achieve spatial diversity in MIMO systems. In Alamouti space time block coding (STBC), every two continuous transmit symbols, s_1 and s_2, are coded into a transmit symbol matrix as follows:

$$s = \begin{bmatrix} s_1 & -s_2^* \\ s_2 & s_1^* \end{bmatrix} \qquad (3)$$

Where (.)* indicates the complex conjugate operation, and s is the space-time code with its columns representing space dimension while the rows representing the time dimension [14].
Alamouti encoded signal is transmitted from the two transmit antennas over two symbol periods. During the first symbol period, two symbols s_1 and s_2 are simultaneously transmitted from the two transmit antennas. During the second symbol period, these symbols are transmitted again, where $-s_2^*$ is transmitted from first transmit antenna and $s_1{}^*$ is transmitted from the second transmit antenna. With orthogonal STBC, s_i and $s_i{}^*$ ($i = 1, 2$) have the same PAPR properties, and thus the PAPR reduction needs to be done only for the first symbol period [15].

3. The PAPR of the signal:

The PAPR of the signal, $x(t)$, is then given as the ratio of the peak instantaneous power to the average power, written as

$$PAPR = \frac{\max_{0 \le t \le T} |x(t)|^2}{E\left[|x(t)^2|\right]} \qquad (4)$$

Where $E[\ \cdot\ ]$ is the expectation operator. From the central limit theorem, for large values of N, the real and imaginary values of $x(t)$ becomes Gaussian distributed. The amplitude of the OFDM signal, therefore, has a Rayleigh distribution with zero mean and a variance of N times the variance of one complex sinusoid.

In practice, they calculate the probability of PAPR exceeding a threshold as measurement index to represent the distribution of PAPR. This can be described as "Complementary Cumulative Distribution Function" (CCDF) and its mathematical expression as [16]

$$CCDF(PAPR(x(n))) = P_r(PAPR(x(n))) > PAPR_0 \quad (5)$$

Due to the independence of the N samples, the CCDF of the PAPR of a data block with Nyquist rate sampling is given by

$$P = P_r(PAPR(x(n))) > PAPR_0 = 1-(e^{-PAPR_0})^N \quad (6)$$

This expression assumes that the N time domain signal samples are mutually independent and uncorrelated and it is not accurate for a small number of subcarriers. Therefore, there have been many attempts to derive more accurate distribution of PAPR [16].

There have been many attempts to find the exact probability distribution of PAPR for analysis MIMO-OFDM system PAPR performance is the same as if each SISO single antenna. For the entire system, the PAPR is defined as the maximum among all the PAPR transmit antennas from [17]:

$$PAPR_{MIMO-OFDM} = \max_{i \le i \le n_t} PAPR_i \quad (7)$$

Where PAPRi denotes the PAPR of transmit antenna. Specifically, since in MIMO-OFDM, Mt N time domain samples are considered compared to N in the SISO-OFDM system, the CCDF of PAPR in MIMO-OFDM is written as follows [17]:

$$P_r(PAPR_{MIMO-OFDM} > PAPR_0 = (1-e^{-PAPR_0})^{M_T N} \quad (8)$$

4. Techniques for PAPR reduction:

4.1. Clipping and filtering:

One of the methods used for eliminate this high peaks is clipping and filtering method**.** The OFDM signal is deliberately clipped at a particular threshold value before amplification in this method [18]**.** The large peaks of OFDM signals are arose with a very low probability and hence clipping could be an effective technique for the reduction of the PAPR.

However, clipping cause important in-band distortion and out-of-band noise which will indirectly degrades the bit error rate performance and the spectral efficiency. Direct clipping suppresses the time-domain OFDM signals of which the signal powers exceed a certain threshold. The penalty is the significant increase of out-of-band energy. Peak windowing or filtering after direct clipping can be used to reduce the outof- band energy. After the filtering operation, the peak of the time-domain signal may increase. Hence, recursive clipping and filtering (RCF) can be used to suppress both the out-of-band energy and the PAPR. RCF can be modified by restricting the region of distortion to obtain improved error performance, filtering is done after clipping in order to eliminate unwanted frequencies caused by the clipping process [19].
Clipping and filtering algorithm for OFDM transmitter block diagram as shown in figure 2.

The main idea of Clipping and Filtering algorithm is to limit distortion of the frequency domain to approximate estimate and processing. The processing steps are as follows:

a) The frequency domain signal through the IFFT transform, received oversampling time-domain signal.
b) Clipping in the time domain, and then clipping distortion to the frequency domain by FFT transform.
c) Out-of-band signal is set to zero, artificially.
d) Using IFFT converted to time domain signal, and output.

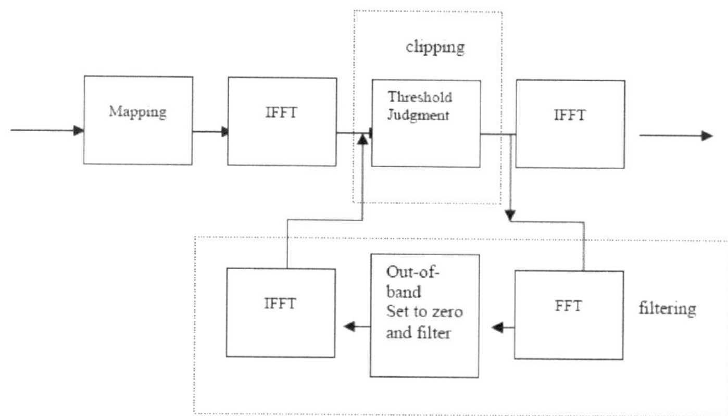

Figure.2 Clipping and filtering algorithm for OFDM transmitter block diagram

4.2. Selective mapping method (SLM):

Another solution is to use selective mapping method (SLM): The entire data stream is divided into different blocks of N symbols each. Each block is multiplied with U different phase factors to generate U modified blocks before giving to IFFT block. Each modified block is given to different IFFT block to generate OFDM symbols. PAPR is calculated for each modified block and select the block which is having minimum PAPR ratio. This technique can reduce PAPR considerably. But this technique will increase circuit complexity since it contains several IFFT calculations [1].

Let.s define data stream after serial to parallel conversion as X= $[X_n = X_0, X_1, ...X_{N-1}]^T$.

Initially each input $Xn^{(u)}$ can be defined as equation

$$x_n^{(n)} = x_n.b_n^{(n)} \qquad (9)$$

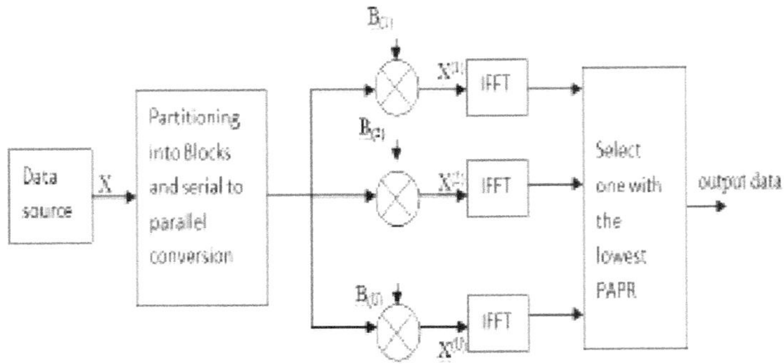

Figure 3. Block Diagram of OFDM transmitter with the SLM Technique [1]

$B^{(u)}$ can be written a $x_n^{(u)} = \left[x_0^{(u)} x_1^{(u)}, \ldots\ldots\ldots x_{N-2}^{(u)}\right]^T$. Where n= 0, 1, 2.........N-1, And u=0,1,2...U to make the U phase rotated OFDM data blocks. All U phase rotated OFDM data blocks represented the same information as the unmodified OFDM data block provided that the phase sequence is known [20].

After applying the SLM technique, the complex envelope of the transmitted OFDM signal becomes:

$$x(t) = \frac{1}{\sqrt{N}} \sum_{n=0}^{N-1} x_n e^{j2\pi n \Delta f t}, 0 \le t \le NT \qquad (10)$$

Where $\Delta f = \dfrac{1}{NT}$, NT is the duration of an OFDM block.

Output data of the lowest PAPR is selected to transmit PAPR reduction effect will be better as the copy block number U is increased. SLM method effectively reduces PAPR without any signal distortion. But it has higher system complexity and computational burden. This complexity can less by reducing the number of IFFT block [21, 22, 23].

4.3. Partial transmit sequence technique (PTS):

In PTS technique, an input data block of N symbols is partitioned into disjoint subblocks. The subcarriers in each subblock are multiplied by a phase factor.

Figure 4 [24] is the block diagram of PTS technique. From the left side of diagram, the data information in frequency domain X is separated into V non-overlapping sub-blocks and each subblock vectors has the same size N. So for each and every sub-block it contains N/V nonzero elements and set the rest part to zero. Assume that these sub-blocks have the same size and no gap between each other [25]. The sub-block vector is given by

$$X = \sum_{v}^{V} b_v X_v \qquad (11)$$

Where $b_v = e^{j\phi_v} (\phi_v \in [0, 2\pi])$ { $v = 1, 2, \ldots\ldots, X_v$ }

is a weighting factor been used for phase rotation.

The signal in time domain is obtained by applying IFFT operation on, that is

$$\hat{x} = IFFT(X) = \sum_{v=1}^{V} b_v IFFT(X_v) = \sum_{v=1}^{V} b_v X_v \qquad (12)$$

For the optimum result one of the suitable factor from combination b = [b_1, b_2,.., b_v] is selected and the combination is given by

$$b = [b_1, b_2, \ldots\ldots, b_v] = \arg\min_{(b_1, b_2 \ldots\ldots, b_v)} (\max_{1 \le n \le N} \left| \sum_{v=1}^{V} b_v X_v \right|^2) \qquad (13)$$

Where arg min [(·)] is the condition that minimize the output value of function.

The phase factors are selected such that the PAPR of the subblocks is minimized. Each of the subblocks having the minimum PAPR and hence the combined signal of the different subblocks is having the minimized PAPR.

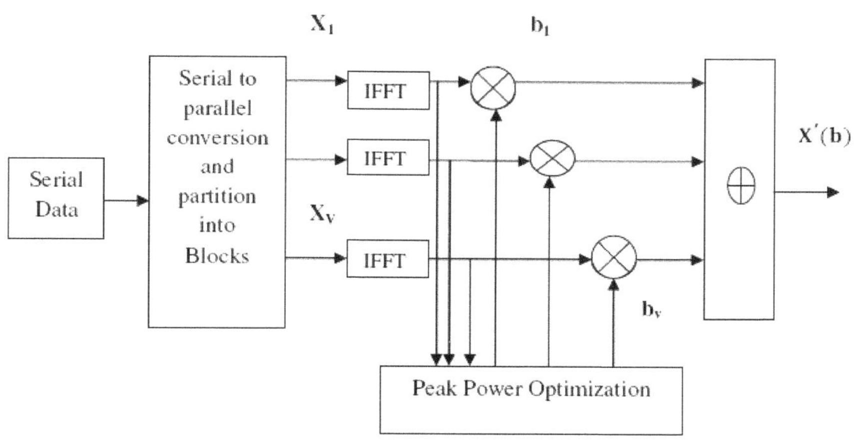

Figure 4. The Block diagram of PTS Technique [21]

5. Simulation analyses of the PAPR reduction:

To implement active set approach for PAPR reduction signal we generate OFDM symbols of length 512 and1024 samples with 301and 601 used subcarriers respectively. Techniques have been done using MATLAB 7.10. The simulation parameters considered for this analysis are summarized in Table 1.

Parameters	Values
System Sub-carrier	512,1024
MIMO Scheme (Tx ×Rx)	(2×2)
Oversampling factor (L)	6
Modulation scheme	QPSK
Phase factor	1,-1, j, -j
Clipping Ratio (CR)	4

Route numbers used in SLM method	M =8
Number of sub-blocks used in PTS methods	V=8
Random OFDM symbols generated	1000

Table 1. Parameters used in clipping and filtering, PTS, SLM and algorithm

Table 1 show the parameters of OFDM signal which is used for PAPR reduction. Here, the number of sub-carriers used are N=512, 1024, An MIMO-OFDM system modulated with QPSK is used for the simulation. The complementary cumulative distribution function (CCDF) of the PAPR for the transmitted signal are plotted after each methods reducing PAPA used.
The flow charts used for PAPR reduction technique is given in Figure 5.

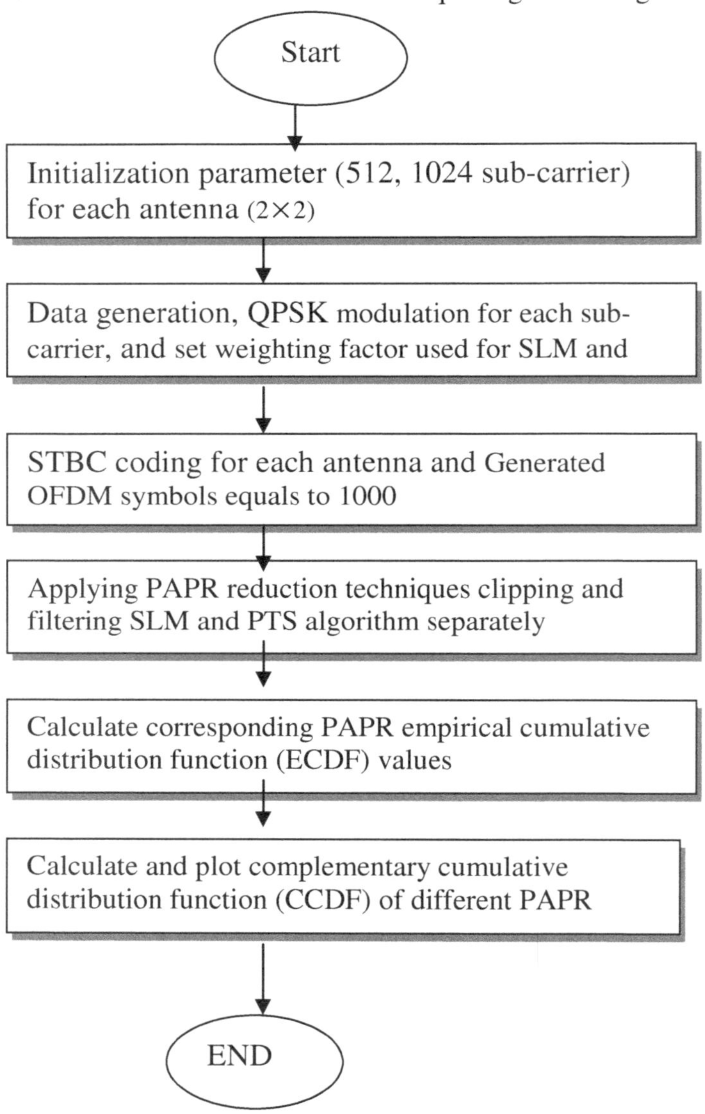

Figure 5. PAPR Reduction algorithm techniques clipping and filtering SLM and PTS

5. Simulation Results:

The results of the simulations are presented in this section. To implement the PAPR reduction of a signal MIMO-OFDM(STBC) we generated OFDM symbols of length 512and 1024 samples 301and 601 tones are used for data transmission and PAPR reduction. Each of the data carrying tones uses a QPSK modulation, an oversampling factor of L = 6 is applied and in our simulation 1000 randomly selected tone sets are generated.

The results of PAR reduction in the simulations are presented as the Complementary Cumulative Density Function (CCDF) of the PAPR of the STBC MIMO-OFDM signals.

The Figures 6,7,8 shows the magnitude of the peak reduction symbol of length 512 with different methods clipping and filtering, partial transmit and select mapping respectively for a system (2x2) MIMO-OFDM (STBC) . We see that the reduction in clipping and filtering is nearly 4 dB, while in some literatures art clipping and filtering achieved only a PAPR reduction of 2dB. This confirms that the STBC coding helps in ways beneficial to reduce fluctuations in the envelope of the MIMO-OFDM with association generally block codes space-system; it takes advantage of the spatial diversity obtained by spatially separated antennas.

In the others methods PTS and SLM reduction of PAPR does not exceed 1.5dB and 2dB respectively

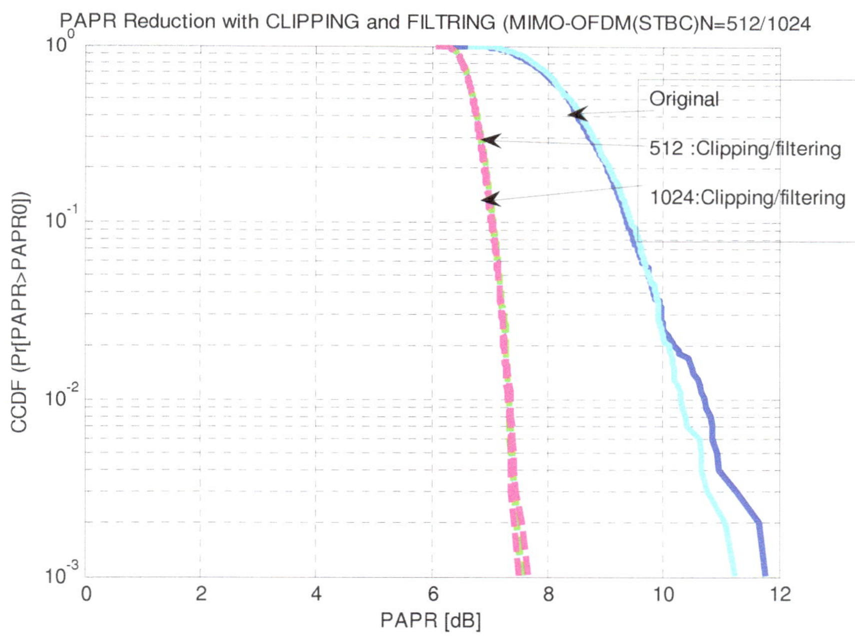

Figure6. PAPR Reduction with clipping and filtering STBC MIMO-OFDM (N=512/1024).

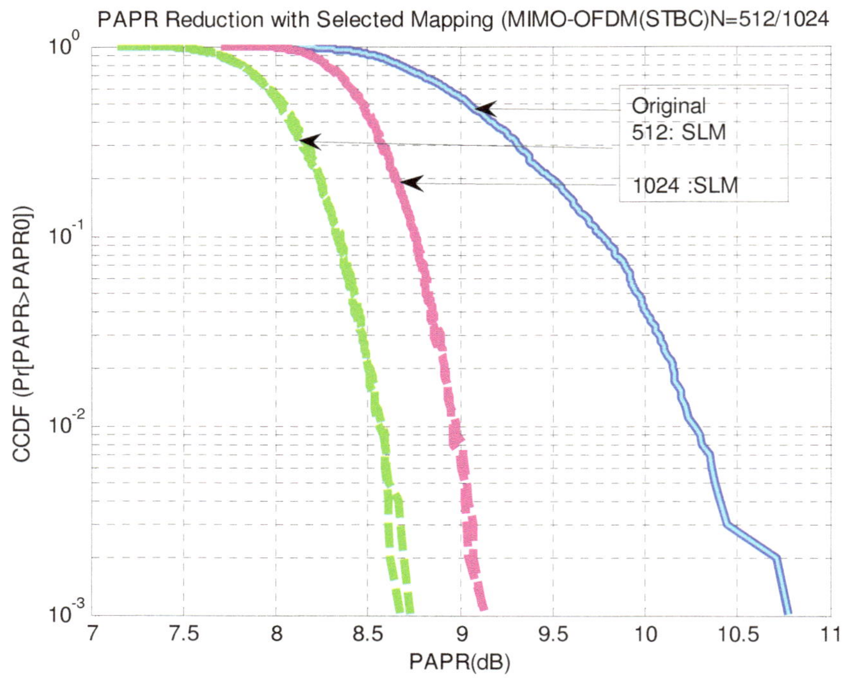

Figure7. PAPR Reduction with selected mapping STBC MIMO-OFDM (N=512/1024)), M=8

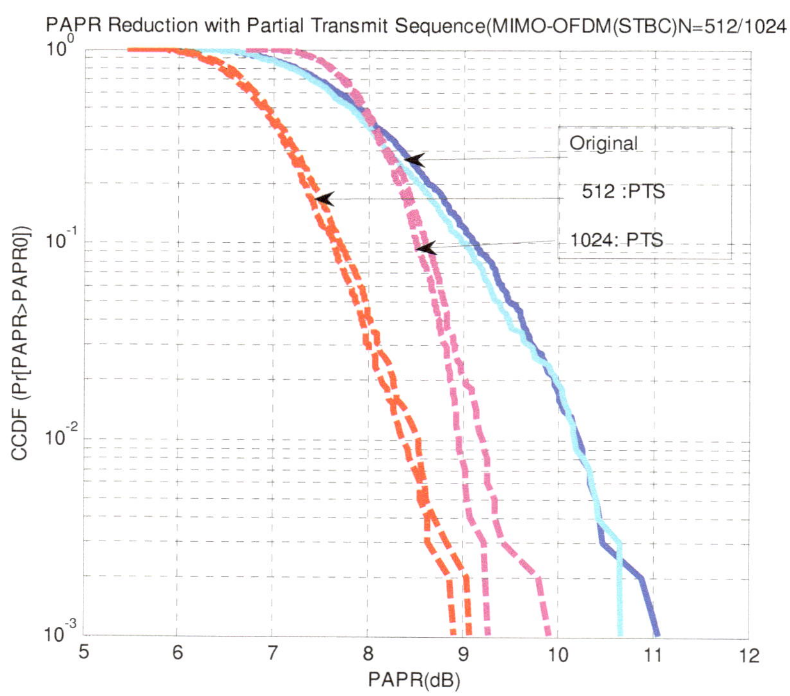

Figure8. PAPR Reduction with Partial Transmit Sequence STBC MIMO-OFDM (N=512/1024), V=8

In the second part a performance comparison between different methods of the PAPR reductions is done to distinguish which technique conserve the reduction found in the transmission

Figure 9,10 shows the performance in terms of CCDF of PAPR of the signal received, a comparison between different methods of the PAPR reductions is done.While the clipping and filtering method shows a slight decrease in PAPR, PTS gives results which diverges from original curves bat with SLM technique PAPR reduction result exceed 2 dB it is equivalent to reduction PAPR calculate in the transmission part. SLM not only reduces the complexity at the reception, but it also reduces the PAPR of the OFDM signal [22]

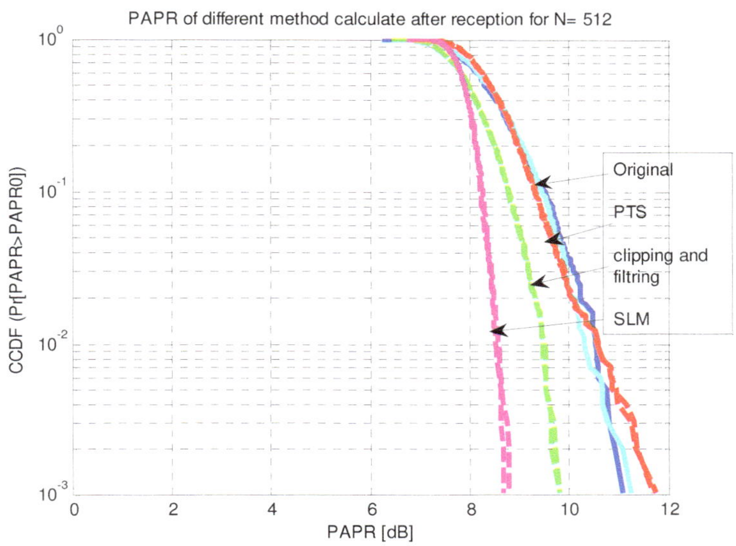

Figure9. PAPR of different methods STBC MIMO-OFDM (N=512), V=8, M=8

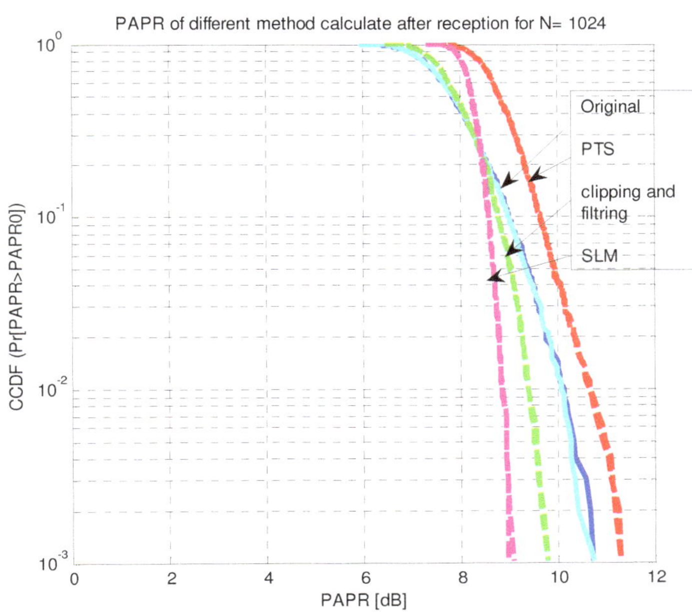

Figure10. PAPR of different methods STBC MIMO-OFDM (N=1024), V=8, M=8

6. CONCLUSION:

In this paper, we present in the first part an analysis of the PAPR reduction method which are clipping and filtering, PTS, SLM in STBC MIMO-OFDM system. Simulation results have shown that clipping and filtering technique gives a better reduction PAPR (4 dB) compared to the others methods. In the second part a result represent by curves CCDF of PAPR of signal received into different methods exhibit a conservation of the PAPR reduction in SLM technique.

7. REFERENCES:

[1] Ms. V. B. Malode1, Dr. B. P. Patil2, "PAPR Reduction Using Modified Selective Mapping Technique," Vol.02, No.2, pp. 626-630 (2010).

[2] S. H. Han and J. H. Lee, "An overview of peak-to-average power ratio reduction techniques for multicarrier transmission," *IEEE Wireless Commun.*, Vol. 12, pp. 56–65, April. 2005

[3] T. Jiang and Y. Wu, "An overview: Peak-to-average power ratio reduction techniques for OFDM signals," *IEEE Trans. Broadcast.*, Vol. 54, pp. 257–268, June. 2008.

[4] M.-S. Beak, M.J. Kim, Y.-H.You, and H.-K. Song, "Semi-blind estimation and PAR reduction for MIMO-OFDM system with multiple antennas," *IEEE Trans. Broadcast*, Vol. 50, pp. 414-424, Dec. 2004.

[5] R. F. H. Fischer and C. Siegl, "Performance of peak-to-average power ratio reduction in single- and multi-antenna OFDM via directed selected mapping," *IEEE Trans. Commun.*, Vol. 57, pp. 3205–3208, Nov. 2009.

[6] Y. Lou J. Palicot, "A classification of methods for efficient power amplification of signals," *Annals of Telecommunications*, Vol. 63, No. 7-8, pp. 351–368 (2008).

[5] X.Li and L.J.Cimini, "Effects of Clipping and filtering on the Performance of OFDM," *IEEE Communications Letters*, Vol.2, No. 5, pp.131-133, May 1998.

[7] M. Mroué, Amor Nafkha,1 Jacques Palicot,1 Benjamin Gavalda,2 and Nelly dagorne2, Performance and Implementation Evaluation of TR PAPR Reduction Methods for DVB-T2, " International Journal of Digital Multimedia Broadcasting, August 2010.

[8] J. A. Davis and J. Jedwab, "Peak-to-mean power control in OFDM, Golay complementary sequences, and Reed-Muller codes," *IEEE Transactions on Information Theory*, Vol. 45, No. 7, pp. 2397–2417, 1999.

[9] S. H. M¨uller and J. B. Huber, "OFDM with reduced peakto-average power ratio by optimum combination of partial transmit sequences," *Electronics Letters*, Vol. 33, No. 5, pp. 368–369, 1997.

[10] R. W. B¨auml, R. F. H. Fischer, and J. B. Huber, "Reducing the peak-to-average power ratio of multicarrier modulation by selected mapping," *Electronics Letters*, Vol. 32, No. 22, pp. 2056–2057, (1996).

[11] Radhia Gharsallah, Ridha Bouallegue "Combined with space time coding multi user detection for CDMA-OFDM/OQAM SYSTEM" International Journal of Wireless & Mobile Networks (IJWMN) Vol.4, No. 4, August 2012

[12] R. W. B. Auml, R. F. H. Fischer, and J. B. Huber, "Reducing the peak-to- average power ratio of multicarrier modulation by selected mapping," *Electron. Lett.*, Vol. 32, No. 22, pp. 2056-2057, Oct. 1996.

[13] X. Li and L. J. Cimini, Jr., "Effect of clipping and filtering on the performance of OFDM," *IEEE Commun. Lett*, Vol. 2, No. 5, pp. 131- 133, May 1998.

[14] Y. A. Khan, M. A. Matin, and S. I. Ferdous, "PAPR Reduction in MIMO-OFDM Systems using Modified PTS and SLM without Side Information," International Journal of Communication Networks and Information Security, Vol. 2, No.3, pp240-247, December 2010.

[15] Yi-Sheng Su, Tsung-Cheng Wu, Chung-Hsuan Wang, and Min-Kuan Chang, "A Low-complexity Cross-Antenna Rotation and Inversion Scheme for PAPR Reduction of STBC MIMO-OFDM

Systems" 16th international workshop on computer aided modeling and design of communication links and networks (CAMAD) IEEE 2011

[16] P. Mukunthan and P.Dananjayan, "Modified PTS with FECs for PAPR Reduction of OFDM Signals," International Journal of Computer Applications, Vol. 11, No.3, pp. 38-43, December 2010.

[17] P.Mukunthan, P.dananjayan, "Modified PTS with FECs for PAPR redaction in MIMO-OFDM system with different subcarriers," international symposium of humanities, Science and engineering Research IEEE 2011

[18] Seung Hee Han and Jae Hong Lee, "Modified selected Mapping Technique for PAPR reduction of coded OFDM signal," IEEE Transaction on broadcasting, Vol. 50, No.3, pp.335-341, Sept 2004.

[19] Pawan Sharma, Seema Verma "PAPR reduction of OFDM signals using selective mapping with turbo codes" International Journal of Wireless & Mobile Networks (IJWMN) Vol. 3, No. 4, August 2011

[20] Yang Jie, Chen Lei, Liu Quan and Chan De, "A Modified selected mapping technique to reduce the Peak to Average Power Ratio of OFDM signal," IEEE transaction on consumer Electronics, Vol53, No.3, pp. 846-851, August 2007.

[21] Stefan H.Muller and Johannes B. Huber,"A Comparison of Peak Power Reduction Schemes for OFDM," In Proc. of The IEEE Global Telecommunications conference GLOBECOM. 97, Phonix, Arizona, USA, pp.1-5, Nov. 1997.

[22] Marco Breiling ,Stefan H. Muller-Weinfurtner and Johanes B.Huber, "SLM Peak-Power Reduction Without Explicit Side Information", IEEE Communications Letters,Vol. 5, No.6, pp.239-241, JUNE 2001.

[23] Jayalath, A.D.S, Tellainbura, C, "Side Information in PAR Reduced PTS-OFDM Signals," Proceedings 14th IEEE Conference on Personal, Indoor and Mobile Radio Communications, Vol.1, Sept 2003.

[24] Oh-Ju Kwon and Yeong-Ho Ha, "Multi-carrier PAP reduction method using sub-optimal PTS with threshold," IEEE Transactions on Broadcasting, vol. 49, June 2003.

[25] Suverna Sengar1, Partha Pratim Bhattacharya, "Performance Improvement in OFDM system by papr reduction", Signal & Image Processing: An International Journal (SIPIJ) Vol.3, No.2, April 2012.

THROUGHPUT PERFORMANCE OF 2×2 MIMO LTE DOWNLINK IN A SPATIAL CORRELATION BASED MICROCELLULAR CHANNEL FOR WIRELESS BROADBAND NETWORKS

Sunil Joshi[1], Deepak Gupta[2], P.C. Bapna[1], Neha Kothari[1], Rashmi Suthar[1]
[1]Department of Electronics & Communication, CTE, MPUAT, Udaipur, India,
sunil joshi7@rediffmail.com
[2]Department of Electronics, V.B. Polytechnic College, Udaipur, INDIA,
simptech@dataone.in

ABSTRACT

Multiple Input Multiple Output (MIMO) technology is going to be a viable alternative for future generation wireless broadband services in order to meet the striving requirements for throughput and system robustness. In Long Term Evolution (LTE), MIMO technologies have been broadly used to get better downlink peak rate, cell coverage, as well as average cell throughput. In the present paper a 2x2 MIMO is taken as baseline configuration for a LTE downlink under a Microcellular propagation scenario considering a non physical correlation based channel with Poor and rich scattering environment. The throughput capacity of the downlink is obtained for poor and rich scattering environments. Besides, two vital aspects of MIMO technique viz Spatial Multiplexing (SM) and Transmit Diversity (TD) are investigated in order to see their effect on throughput of the system. The effect of parameters like Speed of mobile station, number of Multipath, Rician factor (K) on throughput of such systems is reported and discussed. The investigations reported in this paper helps in estimating the throughput capacity of LTE downlink under SM and TD mode.

KEYWORDS

Spatial Multiplexing, Spatial Diversity, Multiple Input- Multiple Output (MIMO), Space Time Coding, Rician factor.

1. INTRODUCTION

In wireless communications, spectrum is an important resource and hence imposes a high cost on the high data rate transmission. Fortunately, the emergence of multiple antenna system has opened another very resourceful dimension – space, for information transmission in the air. A number of studies have demonstrated that multiple antenna system provides very promising gain in capacity without increasing the power and spectrum [1]. The 3rd Generation Partnership Project (3GPP) recommends specifications of Long Term Evolution (LTE), MIMO technologies have been broadly used to get better downlink peak rate, cell coverage, as well as average cell throughput. The main goals of LTE include improving spectral efficiency in 3G networks, allowing carriers to provide more data and voice services over a given bandwidth, lowering costs, improving services, making use of new spectrum and reframed spectrum opportunities, and better integration with other open standards. LTE has introduced a number of new standards to allow IP based wireless mobile broadband. The hybridization of Multiple input multiple output (MIMO) technologies with LTE has generated features such as spatial multiplexing, transmit diversity, and beam forming for better speed and efficiency to support future broadband data service over wireless links [2,3]. The consequence of mobile speed on the performance of 2X2 LTE Downlink MIMO is established in a study reported by Arne Simonsson [4]. He quantified the performance deprivation with reference to the speed of the mobile. The present paper summarizes the studies undertaken pertaining to 2x2 MIMO for LTE downlink standard. The performance of the down link is analyzed under various multi path fading conditions under microcellular propagation environment. The LTE adopts two major MIMO technologies i.e. Spatial Multiplexing (SM) and Transmit diversity (TD). Spatial multiplexing allows transmitting different streams of data simultaneously on the same downlink resource block(s) this increases the data rate of the user [5]. In Transmit Diversity a single stream of data is assigned to the different layers and coded using Space Time Block Coding (STBC). STBC achieves robustness through temporal diversity by using different subcarriers for the repeated data on each antenna [6]. For the LTE downlink, a 2x2 configuration for MIMO is assumed as baseline configuration, i.e. 2 transmit antennas at the base station and 2 receive antennas at the terminal side. Besides, correlation Matrix, a vital channel parameter, is analyzed for of Non – Physical correlation based wireless channel under Poor and rich scattering environment.

2. 2×2 MIMO CORRELATION CHANNEL THEORY

A non-physical correlation channel model [10,17] is considered to describe MIMO Correlation model, generally site-independent and are mostly used for system design, comparison and testing. Correlation-based modeling refers to the spatial correlation present among multipath components arriving with different angle of arrivals and powers at the receiver. The Spatial Correlation matrix can be represented by Kronecker product technique as given by [7]:

$$R_{Spot} = R_{BS} \otimes R_{MS} = \begin{bmatrix} 1 & \alpha \\ \alpha^* & 1 \end{bmatrix} \otimes \begin{bmatrix} 1 & \beta \\ \beta^* & 1 \end{bmatrix} = \begin{bmatrix} 1 & \beta & \alpha & \alpha\beta \\ \beta^* & 1 & \alpha\beta^* & \alpha \\ \alpha^* & \alpha^*\beta & 1 & \beta \\ \alpha^*\beta^* & \alpha^* & \beta^* & 1 \end{bmatrix}$$

Where,

$R_{BS} = \begin{pmatrix} 1 & \alpha \\ \alpha^* & 1 \end{pmatrix}$ for the base station and, $R_{MS} = \begin{pmatrix} 1 & \beta \\ \beta^* & 1 \end{pmatrix}$ for the Mobile Station. The values of α and β can be selected to represent different types of channels, and often real values in the range from 0-1 are used [7].

3. SYSTEM DESIGN AND PARAMETERS

The schematic diagram of the system is shown in figure 1. The LTE downlink for 2x2 MIMO with two transmit and two receive antenna is designed using Agilent's SystemVue tools. A typical channel bandwidth of 10 MHZ with carrier frequency of 2 GHz is selected with FDD duplex scheme and OFDMA access scheme for downlink. The modulation type supported is 16 QAM, a superior modulation scheme for 2x2 MIMO applications [8, 11]. The Physical Downlink Shared Channel (PDSCH) is used as downlink for data transportation across the LTE radio interface. We have used a correlation channel with finite discrete multipath components which are considered to be uniformly distributed about the transmitter and receiver. Each multipath component is considered uncorrelated and characterize by angle of arrival, angular spread and path gain. Two multipath scenarios, first rural one with poor scattering and second urban microcellular with rich scattering environment are investigated. The antenna-to-antenna spacing at transmitting end is assumed to be 2 λ and 0.5λ at receiving end.

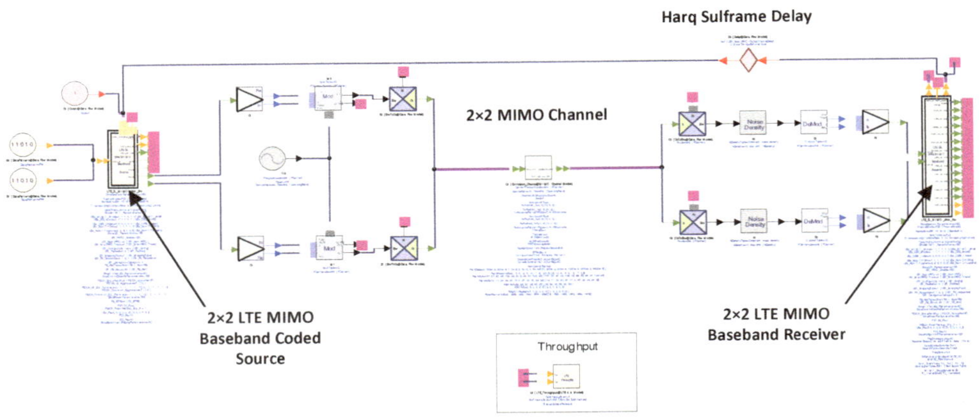

Figure 1. 2×2 MIMO LTE Schematic Design

4. RESULTS & DISCUSSIONS

This model performs the averaged closed-loop HARQ throughput over sub frames, from 'Subframe Start' to 'Subframe Stop' for both PDSCH and PUSCH channel [8]. Each firing, one token is consumed at both 'TBS' port and 'CRC Parity' port. The data input from the 'TBS' port indicates the transport block size for each subframe. If the input value at this port is '0', it is assumed that no transport block is allocated in this subframe. The data input from the 'CRCParity' port is the CRC check result for each subframe, where '1' means CRC check is successful and '0' means CRC check fails[9]

a.) The throughput fraction of the system designed in presence of poor and rich scattering is investigated. A Poor scattering scenario is assumed to have 4 multipath components representing a rural propagation environment and a Rich scattering scenario is assumed to have 12 multipath components representing an Urban microcellular Propagation environment [17,18]. Besides the scenario Rayleigh distribution and rician distribution by varying the values of K factor from 0 to 6 are also investigated to see their effect on throughput performance of the system.

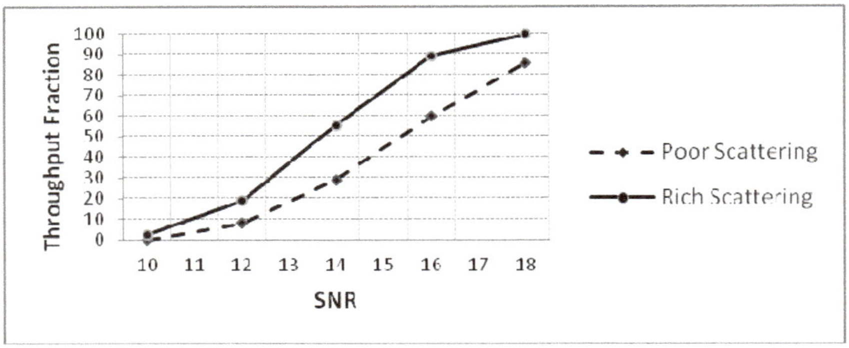

Figure 2.1. Throughput fraction v/s SNR for Spatial Multiplexing (Rayleigh Distribution)

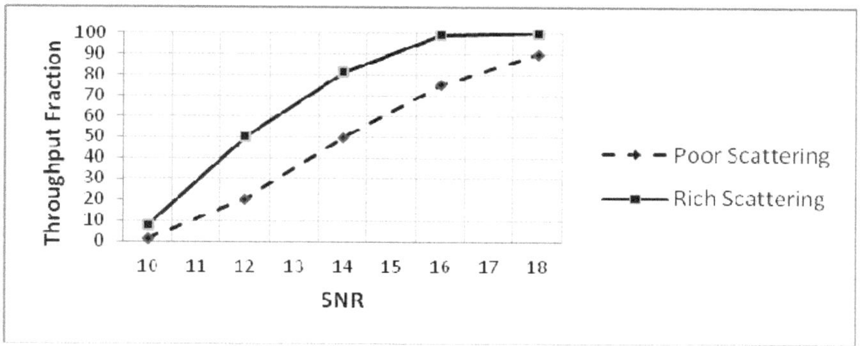

Figure 2.2. Throughput fraction v/s SNR for Spatial Multiplexing (Rician Distribution K=6)

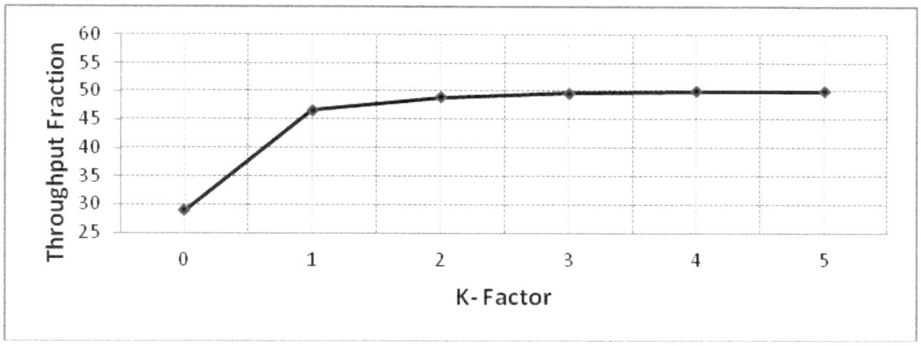

Figure 2.3. Throughput fraction v/s K- Factor for Spatial Multiplexing in poor scattering environment (SNR 14: dB)

The Figure 2.1 reveals that in noisy channel for SNR below 12dB, multipath do not have any significant effect on Throughput fraction. One of the interesting results of SM reveals that the throughput fraction increases almost by two fold when propagation environment is shifted from poor scattering to rich scattering. One of the critical observations derived from fig 2.2 is that under spatial multiplexing mode at relatively low SNR (well below 13dB) the rician factor 'K' is the deciding factor of throughput of the system. Fig 2.3 shows that throughput increases almost linearly when value of K changes from 0 to 1.

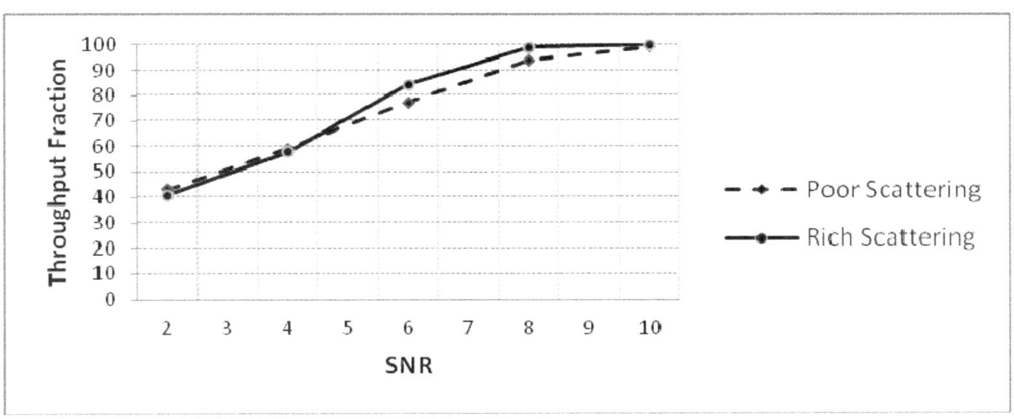

Figure 2.4. SNR v/s Throughput fraction for Transmit Diversity(Rayleigh Distribution)

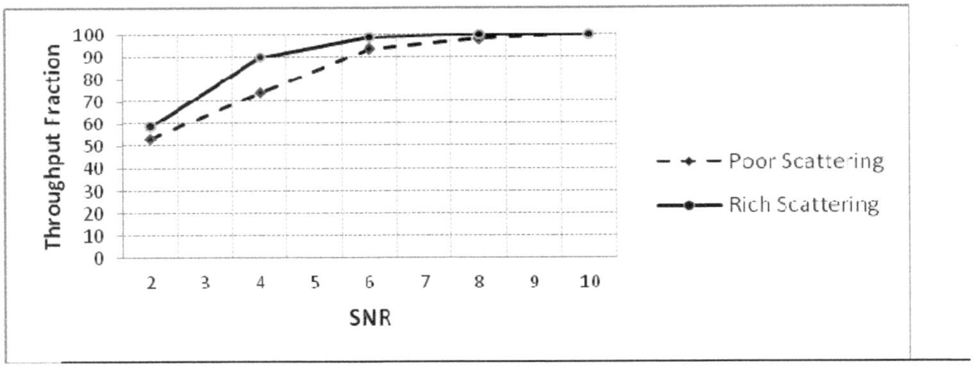

Figure 2.5 SNR v/s Throughput fraction for Transmit Diversity (Rician Distribution K=6)

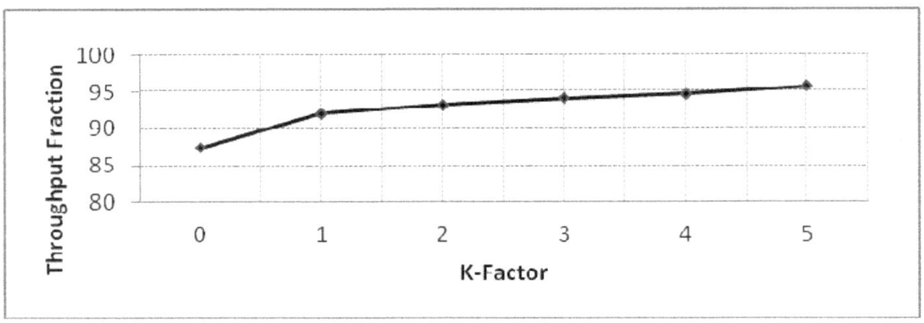

Figure 2.6. Throughput fraction v/s K-factor for Transmit Diversity in poor scattering environment (SNR: 7dB)

Fig 2.4 reveals that in TD mode no. of multipath do not have a significant effect on throughput fraction. As channel conditions improves throughput remains constant (well above 6 dB).Fig 2.5 and 2.6 shows that Presence or absence of LOS component does not affect the throughput fraction.

b.) In mobile environment the vehicle speed is a key factor affecting the data rate of the system. Four vehicle speeds 3, 20, 40, 60Km/hr are separately undertaken to see its effect on throughput performance of the system.

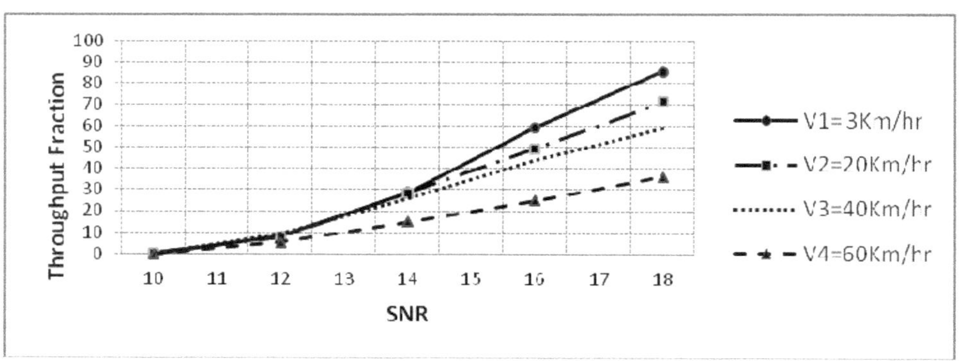

Figure 3.1. SNR v/s Throughput fraction for Spatial Multiplexing with speed variation in Poor scattering environment

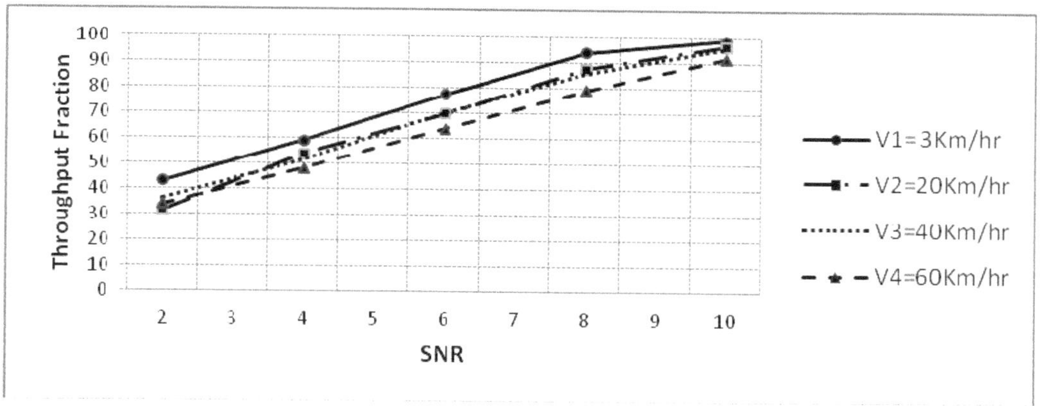

Figure 3.2. SNR v/s Throughput fraction for Transmit Diversity with speed variation in Poor scattering environment

As shown in figure 3.1, in Spatial multiplexing mode, vehicular speed is a key parameter in deciding the throughput of the system. As the speed of Mobile station increases from 20Km/hr to 60Km/hr throughput decreases to about 50%. However, figure 3.2 reveals that in transmit diversity mode mobility of receiver doesn't affect the system throughput.

c) The channel capacity of the system under investigation is sensitive to the multipath environment for different fading distributions. Fig 4.1 and 4.2 indicate channel capacity variations with SNR for Spatial Multiplexing and Transmit Diversity respectively.

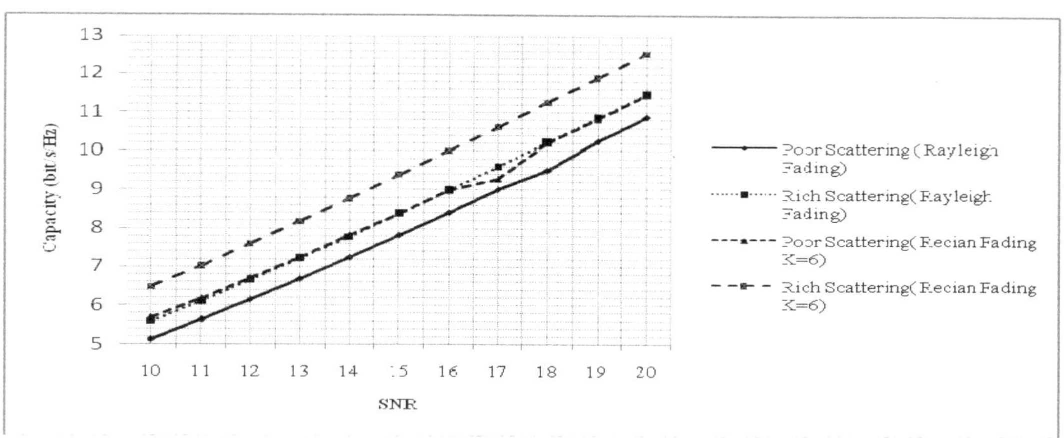

Figure 4.1. Capacity v/s SNR for Spatial Multiplexing

From figure 4.1 it is observed that in spatial multiplexing case the channel capacity obtained is highest in rich scattering environment. A linear variation of capacity with SNR is obtained. In the absence of LOS one can notice that capacity in poor scattering environment is 20% lower than rich scattering for a particular SNR, whereas presence of LOS component enhances the capacity for both the environments up to 17%.

From figure 4.2 it is revealed that in Transmit diversity case also the channel capacity obtained is highest in rich scattering environment. In the absence of LOS one can notice that capacity in poor scattering environment is 14% lower than rich scattering for a particular SNR, whereas presence of LOS component enhances the capacity for both the environments up to 22%.

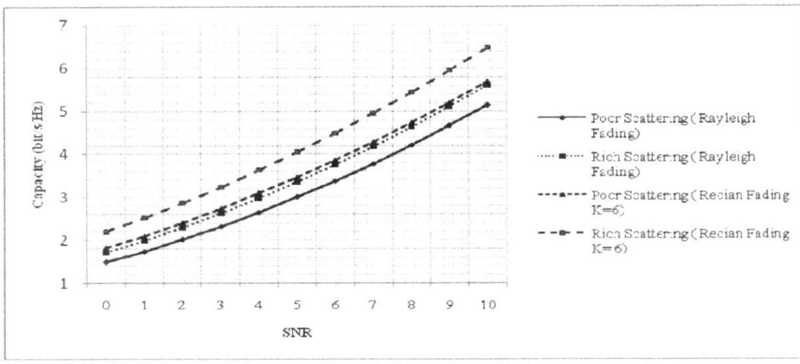

Figure 4.2. Capacity v/s SNR for Transmit Diversity

5. Conclusion

We have investigated the throughput fraction of 3GPP- Long Term Evolution (LTE) downlink system under two popular modes i.e. Spatial Diversity which is commonly used for quality improvement and Spatial Multiplexing which on the other hand is popular in increasing the data rate. These two investigations were in the presence of poor as well as rich scattering channel conditions. The observation were bold and enough that the Presence of rich scattering environment enhances the system performance. In order to gain full advantage of spatial multiplexing the Signal to Noise Ratio (SNR) should be reasonably good while on the other hand transmit Diversity is a better contender for systems operating at relatively low SNR. Through the simulated results it can be very well proved that the region near the edge of the cell where signal strength is relatively low, it is preferable to use diversity based signal stream transmission strategies. As it only requires one spatial channel which is to be significantly stronger than that of the other. In contrast to this the region when the signal strength is relatively high near the base station, it is beneficial to use Spatial multiplexing strategies. We have also envisaged that the relative strength of the direct and scattered component very well expressed by the Rician factor "K" provides an indication of the link Quality. Through our findings and simulated investigation we may easily say that strategies like spatial multiplexing are undesirable for high-speed users whereas Transmit diversity strategy is attractive for high-speed users.

6. Acknowledgement

We thank department of Information Technology (CC&BT Group), Ministry of Communication and Information Technology, Govt. of India, for there financial support for the research project File No.14(10)/2010-CC&BT.

Reference

[1]. Hamid Jafarkhani,"Space-Time Coding Theory And Practice" *University of California, Irvine*

[2]. HSPA to LTE-Advanced, Rysavy Research / 3G Americas, September 2009

[3]. Arne Simonsson, Yu Qian and Jessica Östergaard, "LTE Downlink 2x2 MIMO With Realistic CSI: Overview and Performance Evaluation "978-1-4244-6398-5/10/ ©2010 IEEE

[4]. 3GPP, TR 25.814, "Physical layer aspects for evolved Universal Terrestrial Radio Access (Release 7)".

[5]. Juho Lee, Jin-Kyu Han, and Jianzhong (Charlie) Zhang "MIMO Technologies in 3GPP LTE and LTE-Advanced" 31 May 2009

[6]. Agilent Application note, 5989-8973EN, "MIMO Channel Modeling and Emulation Test Challenges"

[7]. 3GPP, TR25.996, Spatial Channel Model for Multiple Input Multiple Output (MIMO)

[8]. Agilent Application note, 5989-8972EN " Accounting for Antenna and MIMO Channel Effects Using Agilent SystemVue"

[9]. Dr. Essam Sourour "Diversity and MIMO systems for fading channels" Alexendra University, .ppt.

[10]. "A Wideband Spatial Channel Model for System-Wide Simulations," IEEE Transactions on Vehicular Technology, Vol. 56, No. 2, March 2007

[11]. C. C. Martin, J. H. Winters, and N. R. Sollenberger, "Multiple-input multiple-output (MIMO)

[12]. "Radio channel measurements," in *Proceedings of IEEE Vehicular Technology Conference (VTC '00)*, vol. 2, pp. 774–779, Tokyo, Japan, May 2000.

[13]. D. Cheung, C. Prettie, Q. Li, J. Lung, "Ricean *K*-factor in office cubicle environment," *IEEE* 802.11-03/895r1, 11-03-0895-01-000n-ricean-k-factor-in-officecubicle- environment.ppt, Nov. 2003

[14]. T. S. Rappaport,*Wireless Communications, Principles and Practice*, Prentice Hall PTR, Upper Saddle River, NJ, USA, 1996.

[15]. Carmela Cozz, and Brian L. Hughes, "Space Diversity in Presence of Discrete Multipath Fading Channel" IEEE TRANSACTIONS ON COMMUNICATIONS, VOL. 51, NO. 10, OCTOBER 2003

[16]. 3G Americas, "UMTS Evolution from 3GPP Release 7 to Release http://3gamericas.com/PDFs/WP 3GA UMTSRel-8 July07.pdf, July 2007

[17]. J. P. Kermoal, L. Schumacher, P. E. Mogensen, and K. I. Pedersen, *Experimental Investigaion of Correlation Properties of MIMO Radio Channels for Indoor Picocell Scenarios*,

[18]. A. Saleh and R. Valenzuela, "A statistical model for indoor multipath propagation," IEEE J. Select. Areas Commun., vol. 5, no. 2, pp. 128–137, 1987.

9

L-SHAPED SLOT LOADED SEMICIRCULAR PATCH ANTENNA FOR WIDEBAND OPERATION

AMEL BOUFRIOUA

Electronics Department, Technological Sciences Faculty, University Constantine 1, Constantine, Algeria

ABSTRACT

In this paper, a dual frequency resonance antenna is analysed by introducing L-shaped slot in a semi circular patch, different parametric studies have allows and the results in terms of return loss and radiation pattern are given. The results show that dual wide bands are achieved and a better impedance matching for the upper and lower resonance are obtained. Also, it is observed that various antenna parameters are obtained as a function of frequency for different value of slot length and width; it is easy to adjust the upper and the lower band by varying these different antenna parameters. The coaxial feed is used to excite the patch antenna. Theoretical results using Matlab are compared with the simulated results obtained from Ansoft HFSS software and shown to be in good agreement.

KEYWORDS

L-shaped, slot loaded, semicircular patch antenna, wideband operation.

1. INTRODUCTION

Today, the state of the art antenna technology allows the use of different types and models of antennas, depending on the area of application considered. With the rapid development of wireless communications, it is desirable to design small size, low profile and wideband multi-frequency planar antennas. Microstrip antennas are now extensively used in various communication systems due to their compactness, economical efficiency, light weight, low profile and conformability to any structure. The main drawback to implementing these antennas in many applications is their limited bandwidth. However, the most important challenge in microstrip antenna design is to increase the bandwidth and gain [1-14]. Several techniques that can be used to achieve multi-band performances such as multilayer stacked patch, multi resonator and insertion of slots and slits [2] of various shapes and sizes in the patch antennas have been proposed recently [1-14]. When a microstrip patch antenna is loaded with reactive elements such as slots, stubs or shorting pin, it gives tunable or dual frequency antenna characteristics [6]. The most popular technique for obtaining dual-frequency behaviour is to introduce the slots on a single patch [1-4]. Since the slots are cut at an appropriate position inside the patch, they neither increase the patch size nor largely affect the radiation pattern of the patch [11]. These slots can take different shapes like, rectangular or square slot, step slot, tooth-brush shaped slot, V-slot, U-slot, etc [11]. The slot adds another resonant mode near the fundamental mode of the patch and realizes dual frequency response [11].

In this paper, we present a semicircular microstrip patch antenna with L-shaped slot. The proposed antenna can completely cover two bands and provides a significant size reduction. Dual frequency is tuned by changing the dimensions of the slot. In this paper the simulation resultants and the performance analyses using high frequency structure simulator (HFSS) software and Matlab of the proposed semicircular microstrip patch antenna with L–shaped slot are presented, different parametric studies will be allowed and the effect of the various antenna parameters on the return loss and the radiation of the proposed antenna will be presented, also we will present the effect of the substrate on the band width and the return loss with lower and upper resonant frequencies.

2. ANTENNA CONFIGURATION

The configuration of the proposed antenna is shown in Fig. 1. The semi circular microstrip patch of dimensions W × L printed on the grounded substrate, which has a uniform thickness of h and having a relative permittivity ε_r and the dielectric material is assumed to be nonmagnetic with permeability μ_0. The analysis of the half disk patch antenna is similar to that of circular disk patch, but the effective radius changes to 50% reduction in size [6]. The L-shaped slot with dimension (Ls, Ws) is embedded in a semicircular patch (see Figure 2); the L-shaped patch semi-circular antenna features dual-band behaviour.

The patch is fed by a probe coaxial (50Ω) which is easy to fabricate and has spurious radiation [14]. In this feeding technique, the inner conductor of the coaxial connector extends from ground through the substrate and is soldered to the radiating patch, while the outer conductor extends from ground up to substrate.

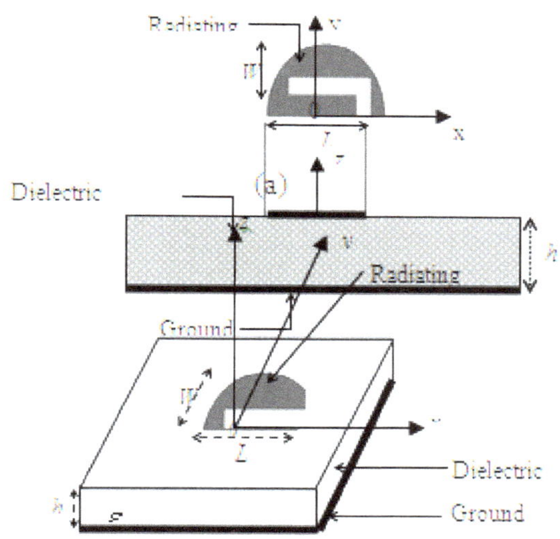

Figure 1. Geometry of L-shaped slot loaded semicircular disk patch antenna; (a) side view, (b) top view

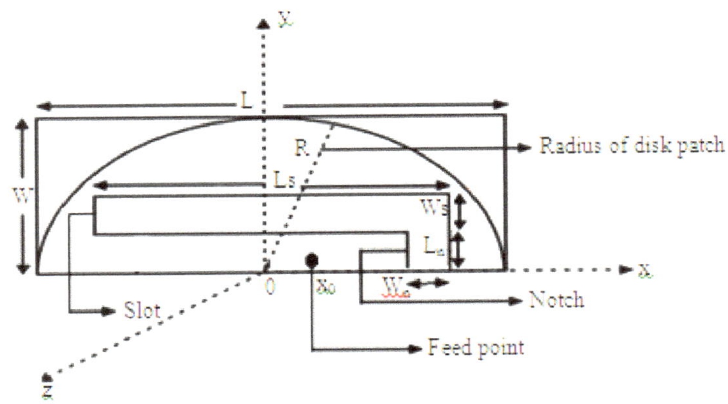

Figure 2. Dimensions of L-shaped slot loaded semicircular disk patch antenna

3. NUMERICAL RESULTS AND DISCUSSION

Table 1 shows the different parameters of this proposed semicircular patch antenna loaded with an L-shaped slot with □r=1. It is worth noting that the feed can be placed at any desired location inside the patch. In this study the feeding is accomplished with a probe coaxial located on the axial of symmetry of the antenna in the point of coordinates x_0 and y_0.

The simulation is done through programs in Matlab and Ansoft HFSS, which is based on Finite Element Method; we compare our results with those available in the reference [6].

Table1. Design parameters of the proposed antenna.

Parameters	R	h	Ls	Ws	Wn	Ln	(x_0,y_0)
Value (mm)	40	15	46.2	6	2.5	6	(12.6, 0)

The return loss is studied in function of frequency; the effect of different physical parameters on the characteristics of the patch antenna is shown. From (figures 3a and b) given by Malab code and HFSS software respectively it is clear that the proposed antenna resonate at two frequencies with two band widths

That is seen from the (Figure3. A), given by Matlab2013 code that the lower resonant frequency fr1=1.48GHz and the upper resonant frequency fr2=2.28GHz. The -10 dB band width of lower and upper resonance frequencies are respectively BP1=14.86%, BP2=10.96%

From (Figure 3. b) obtained from HFSS14.0 software the two resonant frequencies are fr1=1.42GHz; fr2=2.23GHz. The -10 dB band width of the previous lower and upper resonance frequencies are respectively BP1=8.42%, BP2=19.74%.

The reflection coefficients values that are achieved at theses resonant frequencies are lower than -22 dB. These reflection coefficients values suggest that there is good matching at the frequency point below the -10 dB region.

The theoretical results using Matlab are found to be approximately in good agreement with the simulated results obtained with Ansoft HFSS software and with [6].

In the following Figures (4-14), the theoretical results are obtained from Matlab code.

Figures 4 and 5 show respectively the effect of the relative permittivity and the substrate thickness on the return loss and frequency of the proposed L-shaped slot loaded semicircular disk patch antenna. It can be see clearly that the substrate has a stronger effect on both the lower and upper resonant frequencies.

The effect of the radius R of the disk patch on the return loss and frequency is given in Figure 6.

The variation of return loss S11 according to slot length "Ls" is shown by Figure 7, it is observed that the increase of the Ls, increases both the lower and the upper resonance frequencies.

For the variation of S11 according to slot width Ws shown by Figure 8, it is observed that the upper gap of the resonance frequencies increase with increasing value of the slot width. It can be seen clearly that the slot length Ls and slot width Ws have a stronger effect on the lower resonance frequencies than the upper resonant frequencies.

Figures 9 and 10 show the variation of the return loss with frequency for different value of notch length and width respectively, it is observed that the notch length Ln and notch width Wn have a stronger effect on the upper resonance frequencies which increase with the increasing value of the notch length and decrease with the increasing value of the width length than the lower ones which almost remain constant for the variation of these parameters (Ln and Wn).

From Figure 11, it can be observed that when the notch length equal to the slot width, the variation of this two parameters together have a stronger effect on both the lower and upper resonance frequency compared to previous Figures 8 and 9 which show that for the Figure 8, the variation of the slot width has slightly effects the upper resonant frequency and has a stronger effect on the lower resonant frequency, however for Figure 9, it is clear that the variation of the notch length has slightly effects the lower resonant frequency (almost remain constant) and has a stronger effect on the upper resonant frequency.

In Figure 12, it is shown that the feed locations have a stronger effect on both the lower and the upper resonance frequencies; furthermore the obtained results show that the lower resonance frequencies vary more significantly when the x_0 change compared to the upper resonance frequencies as well as the other physical parameters of the proposed antenna.

Radiation pattern of the antenna is shown in Figure 13 and 14 for both upper and lower resonance in both principal planes E and H.

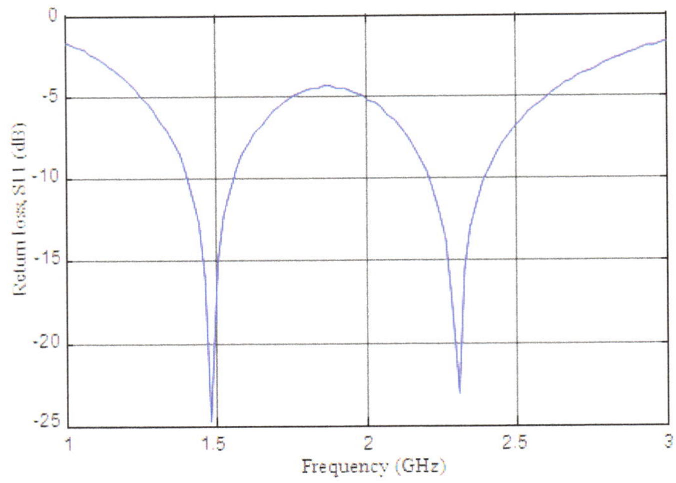

(a)

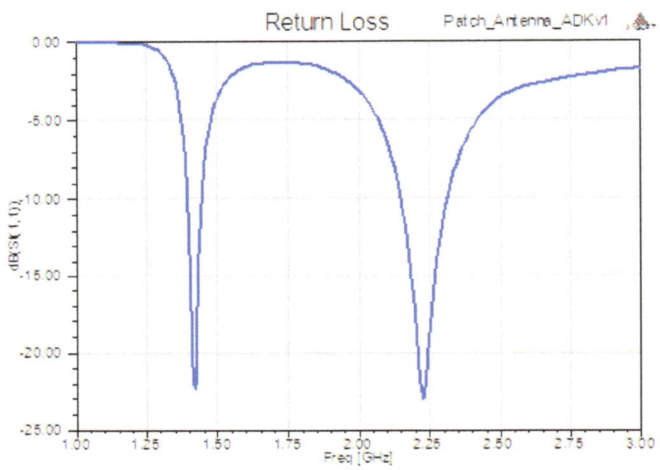

(b)

Figure 3. Comparative plot of return loss with frequency along with theoretical results obtained from Matlab code (a) and simulated results obtained from Ansoft HFSS software (b).

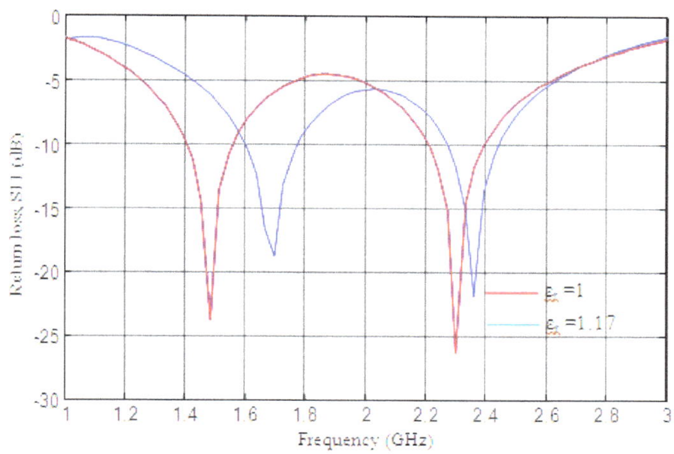

Figure 4. Effect of the relative permittivity on the return loss and frequency of the proposed L-shaped slot loaded semicircular disk patch antenna.

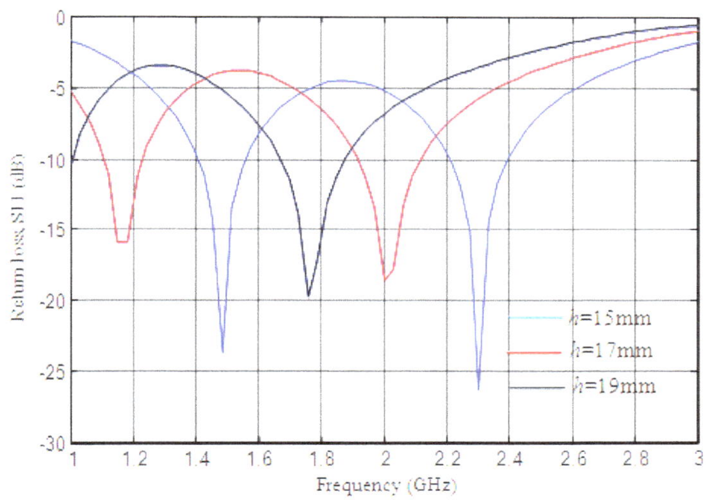

Figure 5. Effect of the substrate thickness on the return loss and frequency of the proposed L-shaped slot loaded semicircular disk patch antenna.

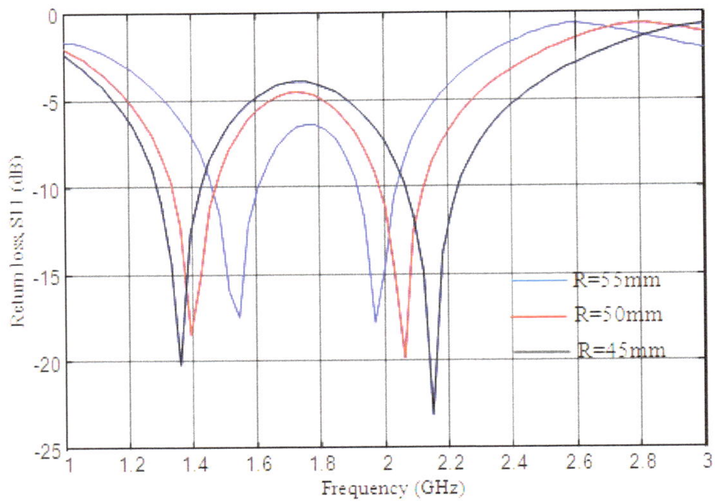

Figure 6. Effect of the radius of disk patch R on the return loss and frequency of the proposed L-shaped slot loaded semicircular disk patch antenna.

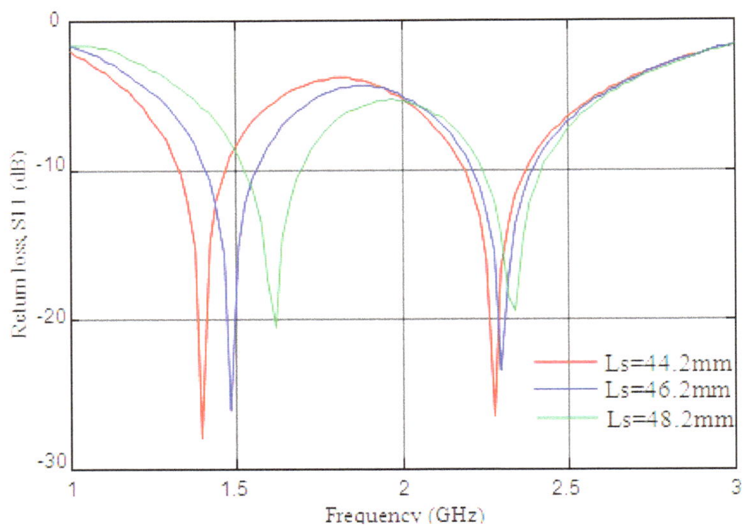

Figure 7. Variation of return loss S11 with frequency obtained from Matlab code for different value of slot length Ls.

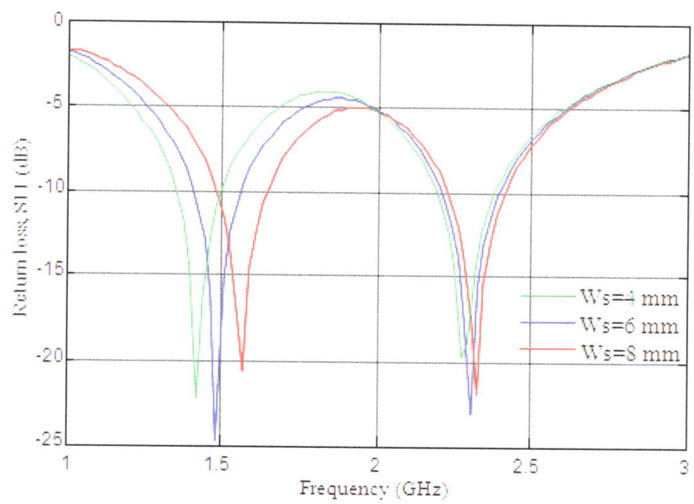

Figure 8. Variation of return loss S11 with frequency obtained from Matlab code for different value of slot width Ws.

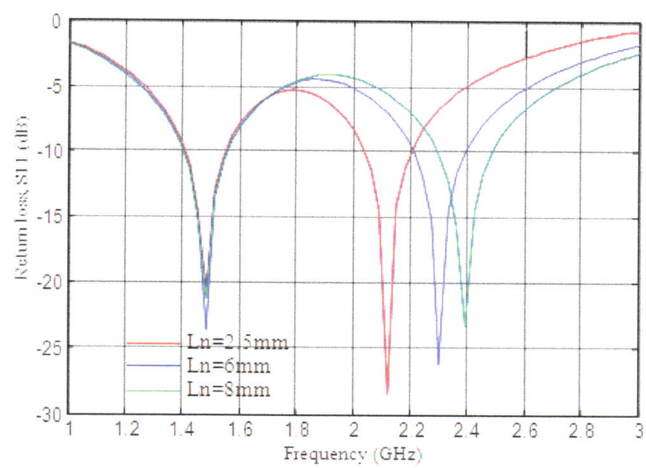

Figure 9. Variation of return loss with frequency for different value of notch length "Ln"

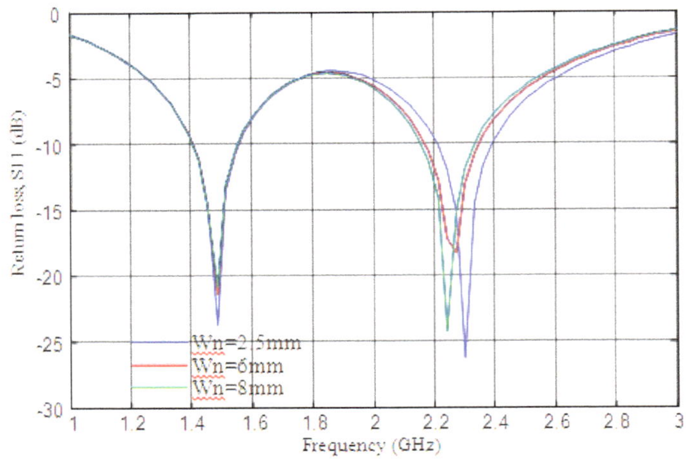

Figure 10. Variation of return loss with frequency for different value of notch width "Wn"

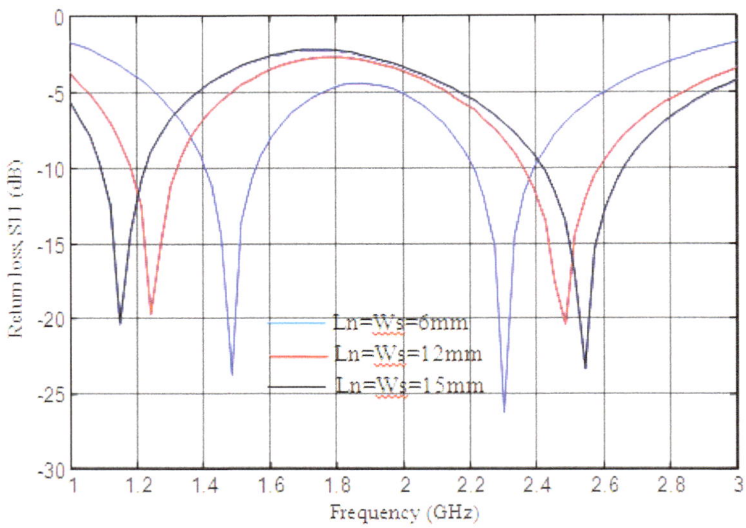

Figure 11. Variation of return loss with frequency for different value of the notch width and length with the notch length equal to the notch width

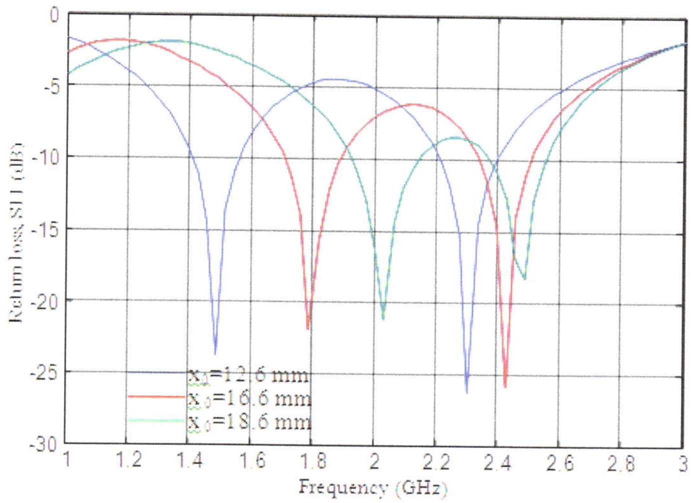

Figure 12. Variation of return loss with frequency for different value of feed point "x_0"

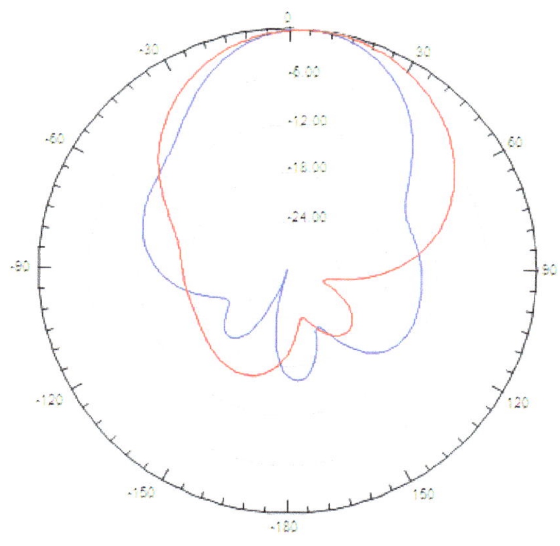

Figure 13. Radiation pattern of L-shaped slot loaded semicircular patch antenna for both upper resonant frequency (blue line) and lower resonant frequency (red line) at E plane

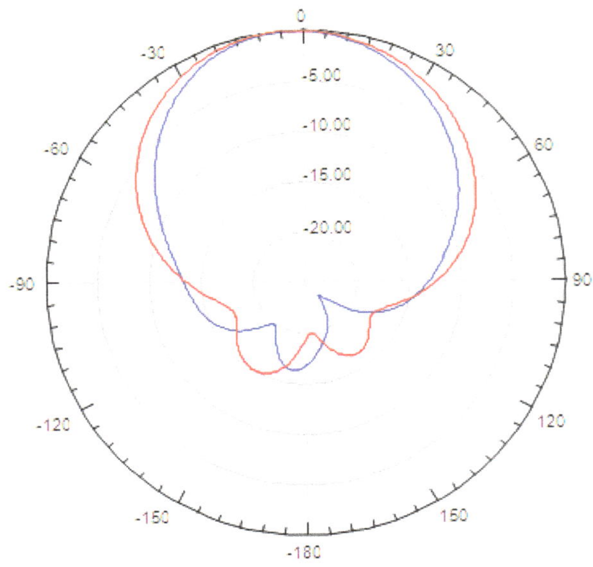

Figure 14. Radiation pattern of L-shaped slot loaded semicircular patch antenna for both upper resonant frequency (blue line) and lower resonant frequency (red line) at H plane

4. Conclusion

It is found that the proposed L-shaped slot loaded semicircular patch structure can operate at two resonance frequencies and consequently this proposed antenna can be used for dual band operation, also the effects of different physical parameters on the characteristics of this structure are investigated, furthermore the effect of the substrate on the band width and the return loss with lower and upper resonant frequencies is given, the proposed structure can be scaled to meet different frequencies of wireless communication systems just by changing the dimension of the main antenna. Numerical results indicate that both the upper and lower resonant frequencies and the band widths depend on the size of the slot dimensions, by properly choosing the location of feed point and the slots two bands can be achieved and controlled. Furthermore, the lower resonant frequencies and band widths are highly dependent on the slot dimensions as well as feed locations, however the upper resonant frequencies and band widths are highly dependent on the notch dimensions. In addition, the radiation pattern of both upper and lower resonant frequencies of the proposed antennas are presented in the principal planes E and H.

References

[1] J. A. Ansari, P. Singh and S. K. Dubey, (2008) "H-shaped stacked patch antenna for dual band operation," Progress In Electromagnetics Research B, Vol. 5, pp. 291–302.

[2] A. A. Razzaqi, M. Mustaqim and B. A. Khawaja, (2013) "Wideband E-Shaped Antenna Design for WLAN Applications," (ICET), IEEE 9th International Conference on Emerging Technologies, 9-10.

[3] S. Maci and G. B. Gentili, (1997) "Dual-Frequency Patch Antennas "IEEE Antennas & Propagation Magazine, Vol. 39, No. 6.

[4] I-F. Chen, C. M. Peng, and S-C. Liang, (2005) "Single Layer Printed Monopole Antenna for Dual ISM-Band Operation," IEEE Trans. On Antennas and Propaga., Vol.53, N. 4. pp. 1270–1273.

[5] J A. Ansari, S. K. Dubey, P. Singh, R. U. Khan, B. R.Vishvakarma, (2008) "Analysis of U-slot loaded patch for dualband operation," International Journal of Microwave and Optical Technology, Vol. 3, pp. 80-84.

[6] J. A. Ansari, A. Mishra, (2011) "Half U-slot loaded semicircular disk patch antenna for GSM mobile phone and optical communications," Progress In Electromagnetics Research C, Vol. 18, pp. 31-4.

[7] D. K. Srivastava, J. P. Saini, D. S. Chauhan, (2009) "Broadband stacked H-shaped patch antenna," International Journal of Recent Trends in Engineering, Vol. 2, pp. 385-389.

[8] E. Wang, J. Zheng, (2009) "A novel dual-band patch antenna for WLAN communication," Progress In Electromagnetics Research C, Vol. 6, pp. 93-102.

[9] A. Boufrioua, A. Benghalia, (2006) "Effects of the resistive patch and the uniaxial anisotropic substrate on the resonant frequency and the scattering radar cross section of a rectangular microstrip antenna" Elsevier, AST, Aerospace Science and Technology, Vol. 10, pp. 217-221.

[10] A. Boufrioua, (2014), "Bilayer microstrip patch antenna with loaded with U and half U-shaped slots," (ICMCS), IEEE International Conference on multimedia computing and systems, 14-16.

[11] A. A. Deshmukh, K. P. Ray, (2010), "Resonant length formulations for dual band slot cut equilateral triangular microstrip antennas," Wireless Engineering and Technology, Vol. 1, pp. 55-63.

[12] H. F. Pues and A. R. Van De Capelle, (1989), "An impedance matching technique for increasing the bandwidth of microstrip antennas," IEEE Trans Antennas Propagat, Vol. 37, pp. 1345-1354.

[13] M. K. Meshram, B. R. Vishvakarma, (2001), "Gap-coupled microstrip array antenna for wide-band operation," International Journal of Electronics, Vol. 88, pp. 1161-1175.

[14] I. J. Bahl and P. Bhartia, (1980) Microstrip antennas, Edited by M. A Dedham, Artech House.

New Strategy to optimize the Performance of Spray and wait Routing Protocol

Mohamed Ababou[1], Rachid Elkouch[2], Mostafa Bellafkih[3] and Nabil Ababou[4]

[1]National Institute of Posts and Telecommunications - INPT, Rabat, Morocco
University of sciences Mohamed 1 -FSO, Oujda, Maroc

Abstract

Delay Tolerant Networks have been (DTN) have been developed to support the irregular connectivity often separate networks. The main routing problem in this type of network is embarrassed by time that is extremely long, since connections are intermittent and opportunistic. Routing protocols must take into account the maximum constraint encountered in this type of environment , use effective strategies regarding the choice of relay nodes and buffer management nodes to improve the delivery of messages and the time of their delivery . This article proposes a new strategy that optimizes the routing Spray and wait. The proposed method uses the information contained in the messages delivered mostly paths traversed by the messages before arriving at their destination and the time when nodes have receive these messages. Simulation results show that the proposed strategy can increase the probability of delivery and minimizing overhead unlike FIFO technology used with the default routing ' sprat and wait'

Keywords

DTN, Store and forward, networks, forwarding strategies, buffer management, routing, epidemic, spray and wait, Queue modes, FIFO.

1. Introduction

The objective of DTN networks [1] is to allow subnet that they know the weak communications in environments where the connection is intermittent, to routes messages between nodes using the principle of store and forward. In this type of network, nodes can forward messages to others nodes when they enter their transmission range. Due to the mobility of nodes, there is no guarantee to find at path between source and destination. To overcome this problem, DTN network uses the mechanism of messages replication to increase the chance that one of the message copies reaches its destination and uses information flowing through the network to choose the . The epidemic routing [2] and spray Wait [3] are among several which sent messages in DTN network using the replication mechanism and PROPHET [4] that uses the network information to choose the relay node .To evaluate the performance of these routing protocols in the absence of real traces, several simulator was developed as the network simulator (NS) and the opportunistic network environment simulator (the one) [5] . The evaluation is based on several metrics, such as the delivery probability, overhead ratio, the delivery rate and the number of hops that have carried messages.

This article proposes a strategy called 'Supp - copiesleft -Tran' to improve the performance of spray and wait routing protocol in terms of delivery probability, overhead ratio and Hop count average. The remaining paper is prearranged as follows .Section 2 elaborates existing forwarding strategies. Section 3 is about routing protocol in DTN networks, Section 4 About mobility models used, Section 5 develops performance metrics, Section 6 the new forwarding strategy, Simulation and results simulates in section 7 by a conclusion at section 8.

2. FORWARDING STRATEGIES

2.1. GRTR

In this strategy the encounters nodes try to calculate the delivery predictability between them and the destination of message P(A,D) and P(B,D) which denotes respectively the delivery predictability [6] that a node A and B have for a destination message . the nodes that has the greater delivery predictability carries a message

2.4. MOFO

This strategy use the number of how many times that message has been sent to others hops and order it according to a descending order, the more the number of the sent message is low, the more the message has a chance of being transmitted [7]. If the buffer is full the strategy deletes messages that have been sent many times. Messages that were sent several times are not sent and the ones that have not been sent or transmitted many times are sent because this increases their chances of reaching their destination.

2.5. Location Based Routing

This strategy uses the coordinates of nodes such as GPS coordinates, to determine the distance between its locations and hops location, and then the message will be forwarded to the one that is closer in the coordinate space than the current custodian to destination.

2.6. First in First out (FIFO)

In FIFO queue mode all messages are arranged according to arrival time and the message which has oldest arrival time will be transmitted first [8].

3. PROTOCOL UNDER OBSERVATION

3.1 Spray and Wait Routing Protocol

Spray and Wait [3] routing protocol has been proposed to reduce the total number of copies sent across the network as is the case of epidemic routing. It uses the principle of epidemic routing but with a different strategy, in the spray phase, the nodes sent L copies to neighboring nodes and in the wait phase each relay nodes carrying the message until they meet the destination. in binary spray and wait version ,each node sends half number of message copies to each node encountered in its path until that they have only one copies where they will wait until they meet destination of the messages.

4. ABOUT MOBILITY MODELS USED

4.1 SPMBM

SPMBM (Shortest Path Map Based Movement Model) [4] is a more realistic model it manages the movement of nodes in the simulation map scenario. It will offer destination coordinates, speeds, wait times, and uses Dijkstra's algorithm to find the shortest path to the destination. It places the nodes in random places but selects a certain destination in the map for all nodes.

4.2. Map Route Movement

Map Route Movement MBM [4], is based on a map of a given region, where nodes can randomly choose their future destination, provided they follow the predefined routes on the map.

5. THE PERFOMANCE METRICES MEASURED

As it was stated in many researches, in order to compare routing strategies, some parameters need to be defined to evaluate their performance. The number of the selected metrics depends on several factors. it will be in relative terms.

5.1. Overhead ratio

Overhead ratio [4] can be defined as the subtraction of delivered (BD) from the bundle carried (BC) over bundles carried ((BC-BD)/BC).The overhead ratio reflects how many unnecessary messages are relayed to deliver one message. It reflects transmission cost in a network. The more the value of overhead is low the more the strategy used is efficient; this leads to a minimization of consumption of the network resources.

5.2. Delivery probability

This metric can be defined as ratio of the messages delivered over messages relayed._The higher values of probability mean that the performance of the algorithm is better.

5.3. Hop count average

It is the mean of the number of hops which participate to relayed message from its source to its destination successfully; higher values mean that the message has consumed many network resources before reaching its destination

6. THE STRATEGY SUPP-COPIESLEFT-TRAN

All nodes are mobile with a low density. There is no connection from end to end, and the movement of nodes affects the delivery of the message. When two nodes meet each other, they exchange the messages that they carry.

6.1. Cleaning the buffer

6.1.1.Phase gathering information.

When two nodes A and B meet, they exchange their two lists namely the list of neighbors and the list of unnecessary messages:

•List of neighbors

$VA = \{ VA_i = Ai , i = 1, ..., n\}$ with n: number of neighbors of A.
$VB = \{ VB_i = B_i , i = 1, ... , m\}$ with m: number of neighbors of B.
VA_i and VB_i are neighbors respectively of A and B.

• List of unnecessary messages.

ListeMAN: List of messages that a node must not accept, it contains messages already arrived at their destinations.
ListeMAN (VAi): The list of messages that the neighbor (A_i) of Node (A) shall not accept.
ListeMAN (A) = { ListeMAN (VA_i),i = 1,..., n } where n : the number of neighbors of A.
ListeMAN (B) = {ListeMAN (VBi), i = 1... m} with m: the number of neighbors of B.

•Creation of ListeMAN (AB)

Creation of a new list that contains the messages to remove, by the union of the two previous lists - that is to say those that emanate from multiple nodes - in order to increase the removal rate of unnecessary messages circulating in the network even if a copy has already arrived successfully to its final destination, so their elimination minimizes the consumption of network resources due to the continuity of their dispersion in the network.

ListeMAN (AB) = {ListeMAN (VA_i)} ∪ {ListeMAN (VB_i)}

•Creating a list of common neighbours

Each node compares the list of neighbours it receives with its own list to find the common neighbor nodes.

$VA = \{VA_i = A_i, i = 1... n\}$ where n is the number of neighbors of A.
$VB = \{VBi = Bi, i = 1... m\}$ with m: number of neighbors of B.
$VA \cap VB = \{N_i, 1 ... n\}$

n: number of neighbors shared between A and B.

✓ if the number of neighbors "n" is even:
$VAB = \{N_i \in (VA \cap VB)\ 1 ...n / 2\}$.
$VBA = \{N_i (\cap\ VA\ VB)\ (2 +n) / 2\} ... n.\}$
✓ if the number of neighbors "n" is odd.
$VAB = \{N_i \in (VA \cap VB)1....(n+1)/2\}$.
$VBA = \{N_i (\cap\ VA\ VB), (n +3)/2 ... n\}$. 6.1.2. Updating the buffer

After creating the list of messages that nodes should not accept in their future contacts and list of common neighbors (VA ∩ VB), the two nodes A and B function like the following:
• They clean their buffers by eliminating unnecessary messages in the list ListeMAN (AB).
• Node A send a list ListeMAN (AB) to its neighbors except those in the list VBA that will be supported by node B.
• Node B send a list ListeMAN (AB) to its neighbors except those in the list VAB that will be dealt by node A.
This implies that the neighbors of two nodes clean their buffers by eliminating unnecessary messages in the ListeMAN list (AB).

With this technique it was possible to increase the rate of messages to eliminate to release buffers of neighboring nodes and increase the probability that a Message previously received by its destination stops circulating in the network.

6.1.3. Transmission of messages

6.1.3.1 Form of messages

Table 1 Information Contained In Message.

Message	S	D	(hop, t_i) = (N_1, t_1) ... (N_n, t_n)	Copies left

TABLE 2 Information Contained In Acknowledgement Message.

Ack	D	S	(K_i, t_i)=(k_1, t_1) ... (K_n, t_n)	(S, t_0) {(N_i, t_i)} (D, t_{n+1})

S: Source D: destination,(hop, t_i) :hop involved in transport of the message. t_i: the time when the message has been transmitted to Ni , Size: the size of the message.Copies left: the number of copies remaining to be transmitted.

(Ki, ti) hop who participated in the transport of acknowledgement messages.
{(S, t_0) {(N_i, t_i)} (D, t_{n+1})} the path that the message took to go from the source to the destination.

6.1.3.2 The Priority List Creation .

Each node classifies messages in its buffer; example node A searches for messages that have as destination node B, it range them in descending order based on their injection in the network and in ascending order based on their size and it is at this level that starts transmission.

- **Method for creating the priority list:**

Node A consults at the buffer messages before deciding:
- If B is the source: it ignores the message.
- If B is the destination: it added to the priority list.
- If B belongs to the hop list who participated in the transport of the message to the moment of contact: it ignores the message.

6.1.3.3 Transmission of Priority List

Once the priority list is created the node classifies the messages in its buffers ordering to the number of copies left before forwarding the messages in this list to node B.

6.1.3.4 Transmission or Non-Transmission of Other Messages.

In the case where the node B is not in the set of hops {(S, D, {(Ni, Ti)}} that is to say, it is neither a source nor a destination, nor the one of the nodes that participated in the transfer of messages,

node A uses the information contained in the list of paths supplied by the set of paths extracted from the list ListeMAN (AB):
{(Sj, t0) {(Ni, ti)} (Dj, tn +1)}: list of paths of messages successfully transmitted to their destination.
{(Dj, tn +1) {(Ki, ti)} (Nj, tn)}: list of acknowledgments paths.
Node A, creates a list of pairs of nodes (Ni, D) called:
Listech (D, Ni) ={vector(frequency, hops average number)} which means that it extracts from all messages paths in list ListeMAN (AB), the number of times(frequency) when the two nodes are involved with delivering messages to their success destination and the average number of hops that separate in each path, then classifies messages in ascending order based on the number of occurrence of nodes Ni and D (frequency) and in descending order of hop number between them, thereafter, it selects from its neighbors which must carry the message. The nodes Ni and D are very close if the number of hops between them is minimal, in this case there is a strong chance that the message reaches its destination without any problems.
Once the secondary list is created the node classifies the messages in its buffers ordering to the number of copies left before forwarding the messages in this list to appropriate node.

7. EXAMPLE

Assume A, B ,C,E,F and D are five intermittently connected mobile nodes and {B,C,F,E} are neighbors of A see Figure 1 and Table 3represents the messages buffered at node A and node B, previous to the transmission and Table 4 shows the lists of messages delivered successfully to their destination by each nodes. It is assumed that each message transmission time is 1 second while total DTN transmission time is 4 seconds.

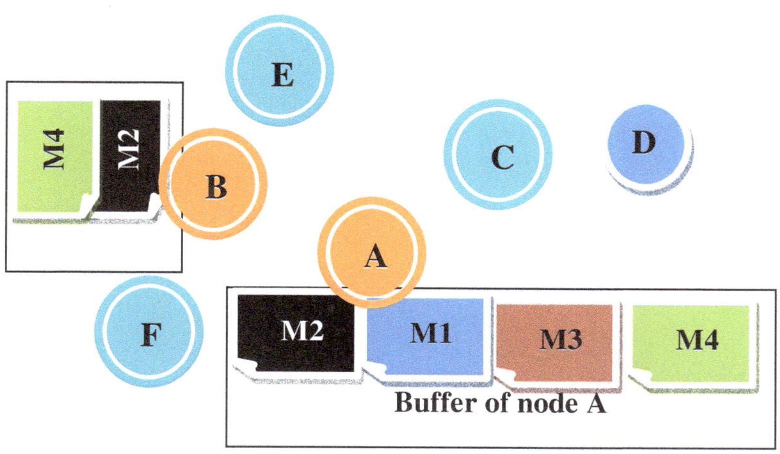

Figure 1 snapshot of encounters nodes before communication between A and B

ListeMAN(A)	ListeMAN(B)	ListeMAN(C)	ListeMAN(E)
M6	M1	M3	M4
M17	M11	M12	
M81	M12	M13	
M19	M13		
M12			

M11
M12
M13

Figure 3 List of messages already deleted by node A , B ,C and E

Buffer of Node A				Buffer of Node B			
Msg	Destination	Copies Left	Arrival Time	Msg	Destination	Copies Left	Arrival Time
M1	D	5	150	M17	D	4	224
M2	D	5	123	M2	D	2	224
M3	D	4	60	M5	D	4	204
M4	D	4	15	M6	D	5	99

Figure 4 snapshot of messages buffered in node A and B before communication

Case 01: Forwarding Sequence with FIFO strategy

Table 4 represents the array of messages at node A and B before the transmission. With FIFO strategy, node A transmits the messages:
Node A forward the message:
{M3,M4} to node B except {M1,M2}, which was already transported them
{M1, M2, M4} to node C except M3 and {M1, M2, M3} to node E except M4
Figure 2 shows the messages buffered at neighbor's node A after transmission. The result shows that the messages M1, M3 and M4 continue to circulate in the network, even if one of its copies was delivered to its final destination. This causes an unnecessary consumption of network resources.

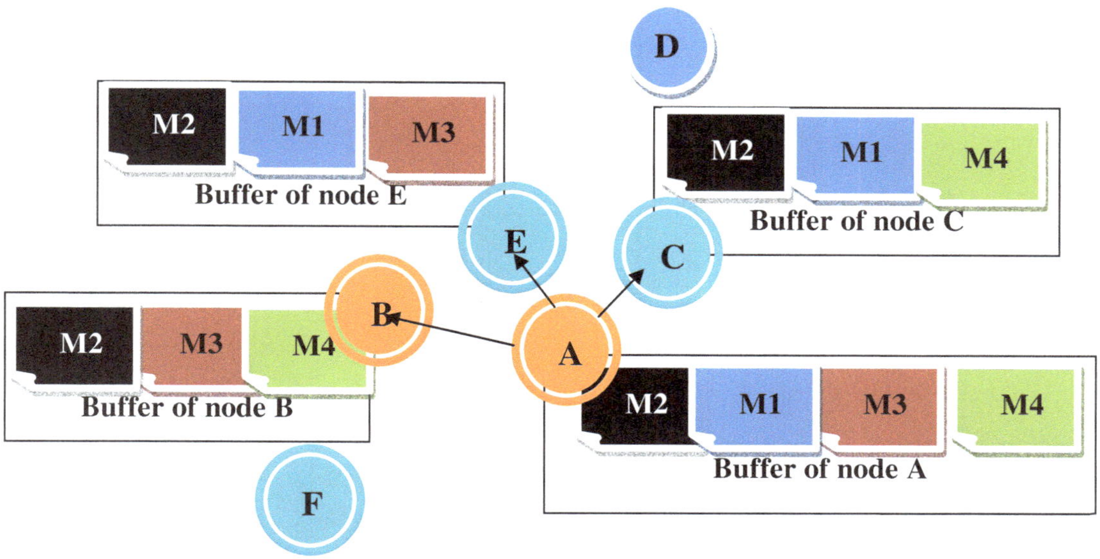

Figure2. Messages buffered in neighbors' node A after transmission with FIFO strategy

Case 02: Forwarding sequence with "Supp-copiesleft-Tran" strategy

Table 5 depicts the organization of messages at node A by Supp-copiesleft-Tran previous to transmission.

With sup-copiesleft-tran strategy, node A follows the following steps before messages forwarding:

Step 1: Node A exchanges first, the list 'LISTMANT' [Table 3], with its neighbors and thereafter, forms the global list that will contain the messages' identifiers of already delivered messages.

ListeMAN(A)	M6	M17	M81	M19	M12	M11	M12	M13
	M1	M11	M12	M13	M3	M12	M13	M4

Table 5. List 'LISTMANT' created by node A in Tran phase

Step 2: Node A updates its buffer and sends the list 'LISTMANT' [Table 5] formed to its neighbors for update their buffer.

Step 3: The node A sends messages that remain in the buffer based on the list of paths 'LISTECH'[Table 6] which means, that after it updated its buffer, node A removes messages that are successfully delivered, that is to say, the messages M1,M3 and M4 , it still has only M2 which will forward it to node C, the more likely to meet node D. Figure 3 shows that the technique avoids the transmission of unnecessary messages if one of its copies arrived at its destination.

Table 6. LISTECH(D,Ni) of node a Created By "Supp-copiesleft-Tran" in Section 6.

pair of nodes	frequency	Average hops
D:C	11,7	0
D:B	5,87	3
D:H	3	3
D:F	3	5
D:E	2	7
D:A	1	1

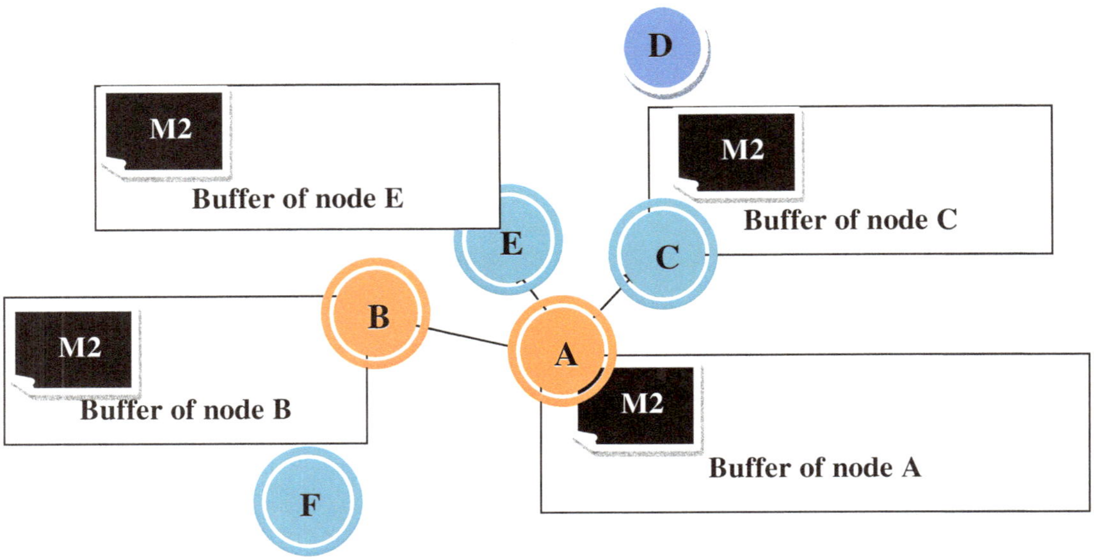

Figure 3 Messages buffered in neighbors' node A after transmission with 'SUP-COPIESLEFT-TRAN' strategy

8. SIMULATION AND RESULTS

8.1 Simulation Environments

This section presents the results of the comparison between the forwarding strategies FIFO and Sup-Tran presented in Section 6, regarding their effectiveness under spray and wait routing protocols. All simulations are done with the ONE (The Opportunistic Network Environment Simulator) [1] written in Java to evaluate the performance of the two methods. The ONE was created by Helsinki University and provided the map of the Helsinki area.

In the simulation, there are two different nodes that may generate and receive a message. One is a pedestrian and the other is a vehicle. The number of two different nodes is 40 and 6 respectively for pedestrian and vehicle. These two types of nodes are moving with different mobility along the map of Helsinki area.

For the pedestrian, the walking speed (i.e., 0.5m/s~1.5m/s) is applied. The moderate speed (i.e., 6m/s~12m/s) is applied to the vehicle. The details of the simulation parameters are shown in Table 7.

Table 7. Simulation parameters

Simulation Parameters	Simulation values
Number of nodes	200 nodes (P: 100, V: 100)
Simulation time	10k-80K for simulation w.r.t time 10K for simulation w.r.t Transmission range
Node movement	Shortest path map based movement, Map Route Movement
Transmission range	10m for simulation w.r.t time 10m-120m for simulation w.r.t Transmission range
Packet transmission speed	250Kbyte/s
Buffer size	10M,50M

8.2. Performance Analysis

To evaluate the proposed method compared with FIFO under spray and wait routing protocol, the simulation has to consider performance metrics such as delivery rate, communication overhead, and number of hops. It is evident that the higher delivery rate means better performance on successful data delivery. However, the effort to get the higher delivery rate, a routing protocol has to send the more data into the networks, these additional data packets may result in communication overhead. Therefore, both delivery rate and communication overhead are analyzed simultaneously.

Finally, average number of hops is compared to demonstrate consumption of resources. Clearly, DTNs accept a tolerable delay for message delivery. On the other hand, some applications do not accept higher number of hops to deliver messages from the source to the destination.

8.3. Simulation Results

All the results got from the simulation are shown below in form of graph representation and observations are discussed.

8.3.1. By Varying Time Intervals

	Spray and wait				
strategy	FIFO	SUP-copiesleft-tran	strategy	FIFO	SUP-copiesleft-tran
created	336	336	delivered	120	133
started	6700	6647	Delivery probability	0.3571	0.3958
relayed	2617	2561	Overhead ratio	20.8083	18.2556
aborted	4081	4084	Latency average	2182.3867	2403.303
dropped	1993	1286	Hop count average	2.8	2111.8
removed	0	759	Hop count average	2.8	2111.8

Table 8. Example of result of simulation with 10000 seconds

Figure 4 shows that increasing the simulation time from 10K to 80K increases the probability of deliverance for both method, which can be explained as more nodes are traveling for a long time, the more they encounter another nodes so they may exchange more messages, but for each time interval, the results show that the strategy Supp-copiesleft-Tran has better performance than the FIFO strategy in terms of messages that can be delivered to their destination.

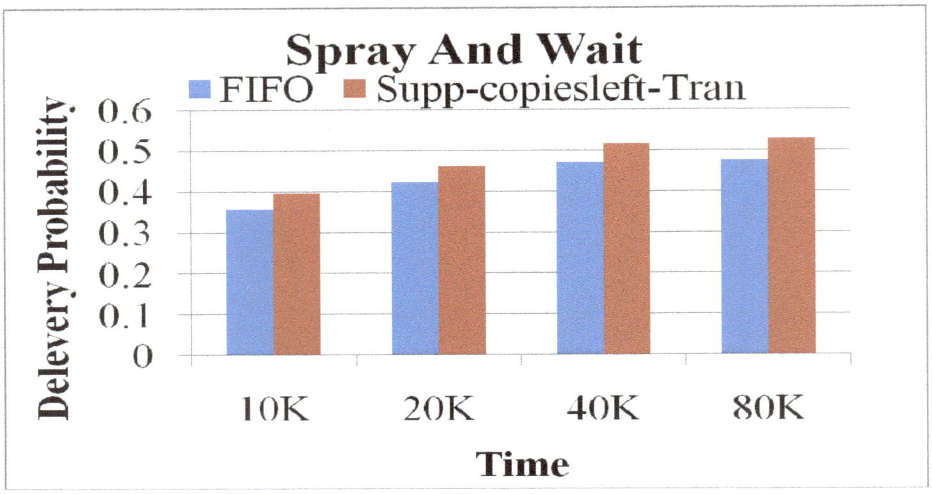

Figure 4. Delivery Probability w.r.t

Figure 5 shows the results of the evaluation in terms of the overhead ratio of FIFO and Supp-copiesleft-Tran by varying the time interval from 10K-80K under spray and wait routing protocol. As shown, the overhead generated by the proposed method is lower than the FIFO strategy in all interval time, which means that with Supp-copiesleft-Tran strategy, the number of messages that continue traveling in the network without reaching their destination. Unlike, with FIFO strategy where this number is higher which means that the messages moving around the network for a long time which causes that the messages consume the network resource.

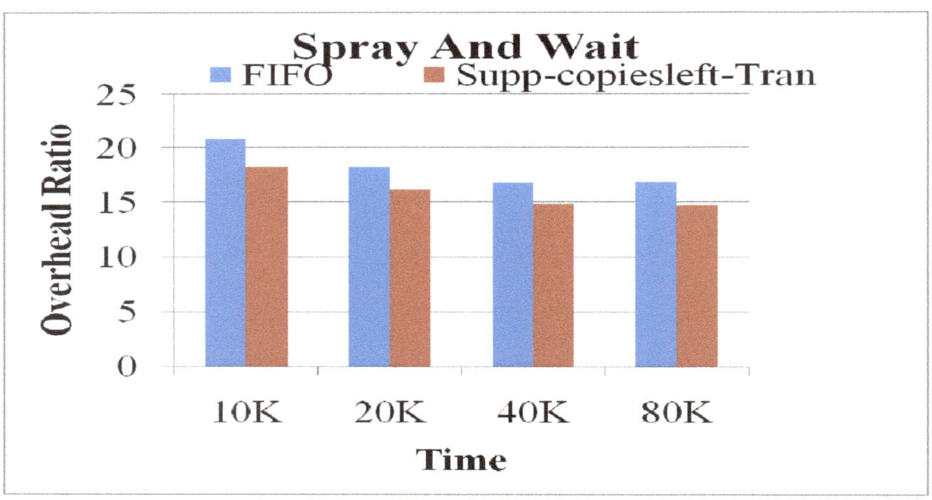

Figure 5. Overhead ratio w.r.t Time

Figure 6 explains the average hops number of the proposed strategy compared to FIFO technique, as the result show, the proposed approach has a smaller pattern of the average hops number. It can see clearly that at various time intervals the hop-count average of "Supp-copiesleft-Tran" is lower than FIFO. The proposed method uses a number of hops slightly greater than the strategy FIFO which is evident because the Supp-copiesleft-Tran method tries to choose the best hops having a higher probability or greater chance to deliver messages, unlike the FIFO technique that transmits the messages based on their arrival time without checking if the relay nodes will encounter destination or not which causes that some nodes carry the messages even if it will never meet the destination, even if it has a small number of hops, it does not mean that is better because the delivery probability is low as its shown above in Figure5.

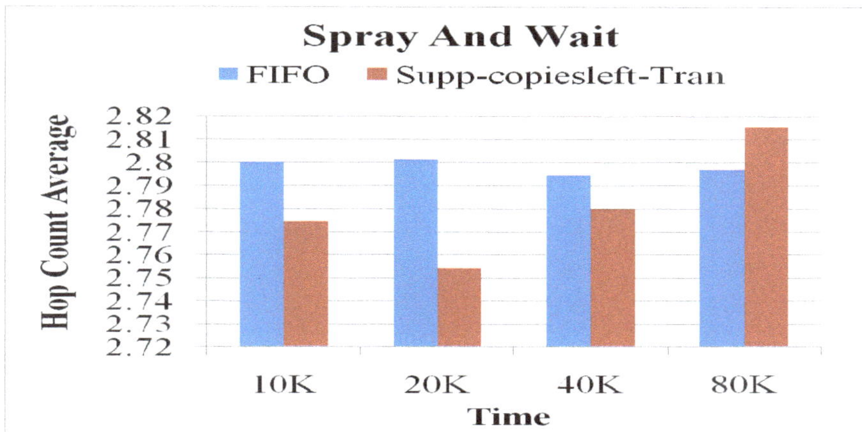

Figure . Hop count average w.r.t Time

8.3.2. By Varying Transmission range

Figure 7 compares FIFO and Supp-copiesleft-Tran strategy in terms of delivery probability by increasing the transmission range. The Supp-copiesleft-Tran strategy proves better delivery probability than FIFO, since increasing the transmission range increases the number of neighbors, which allows nodes using Supp-copiesleft-Tran to deliver the message as close as possible to their destination by choosing the most appropriate hop.

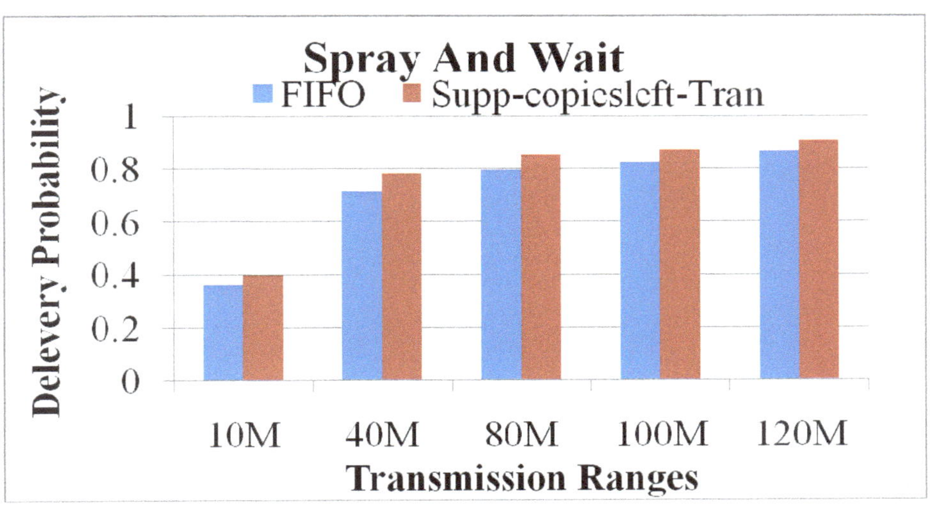

Figure 7. Delivery probability w.r.t Transmission range

Figure 8 shows that in all transmission ranges the overhead of "Supp-copiesleft-Tran" is lower than FIFO strategies. This means that the strategy Supp-copiesleft-Tran is able to choose the best relay to transport messages to the right destination that is reflected in the low value of overhead.

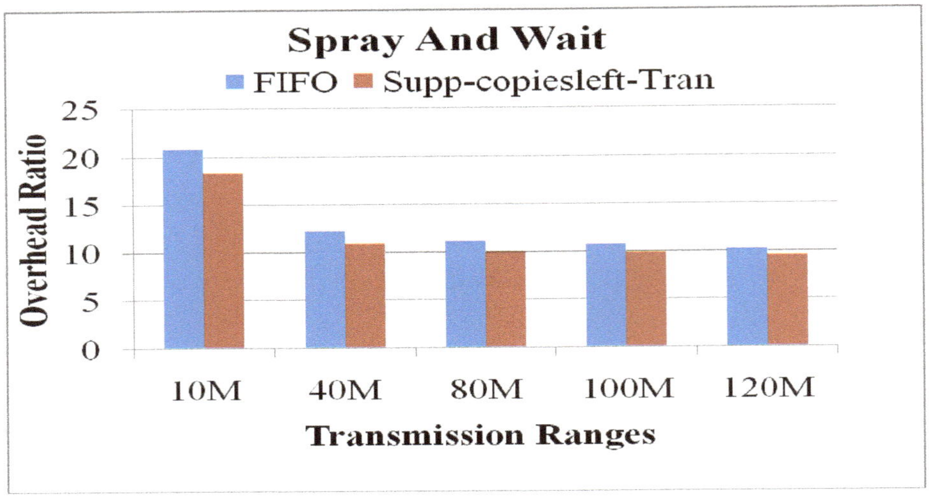

Figure 6. Overhead ration w.r.t Transmission range

Fig 7 presents the impact of transmission range on hop count. It shows that the hops count with FIFO strategy is lower than Supp-copiesleft-Tran. The sup-Tran, especially in the delivery probability and overhead ratio, is more efficient than FIFO.

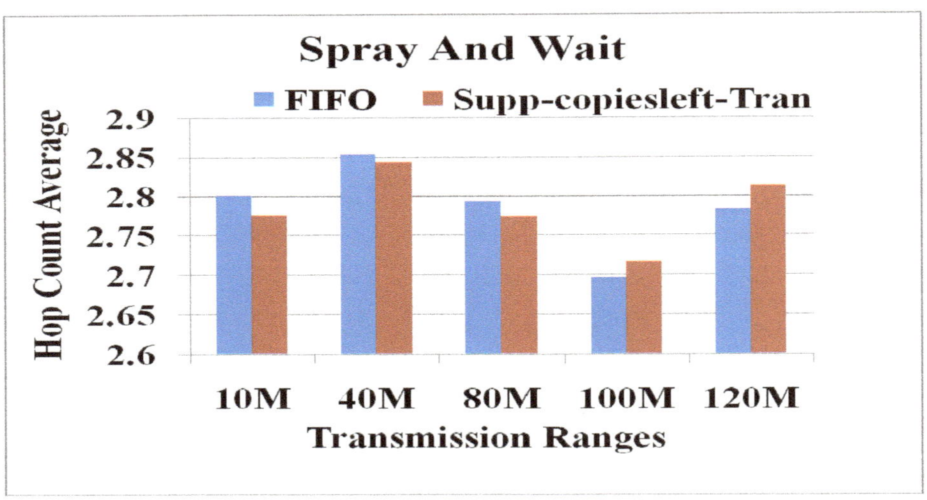

Figure 7. hop count Average w.r.t Transmission range.

CONCLUSIONS

This article proposes a strategy called " Supp-copiesleft -Tran " , which aims to optimize spray and wait routing protocol in delay tolerant networks, to use the network resources in an efficient manner contrary to the FIFO technique, as its name indicates, the proposed technique consists of three phases, the ' Supp' which means that the encountered nodes start to clean their buffer before deciding to transmit messages that are in their buffer. They exchange the list of messages that are successfully delivered to delete the remaining copies that still circulate in the network to release their buffers as well as those of their neighbors. The phase 'copies left' the nodes organize the messages in their buffer in order to the number of messages copies that remaining ,the ones that have not been sent or transmitted many times which increase their chances of reaching their destination.The phase 'Tran' starts with the transmission of the messages selected in phase 'copies left' based on the list of neighbors and the list formed by the paths of messages delivered to calculate the frequency of meeting between destination and neighboring nodes of the two nodes that are in communication as well as the number of hops between them. During the phase "Supp", a proposed strategy optimizes routing sprat and wait by increasing the rate of the removed copies of messages that have been delivered to their destination. Indeed, the list formed by considering the list of the deleted messages of the encountered nodes and their neighbors will be forwarded to the neighbors to clean their buffer. During the phase 'copies lift' the technique give priority or chance to the message that not been sent or transmitted many times, During the " Tran" Phase, the proposed technique optimizes routing sprat and wait in terms of selecting relay nodes that will be carrying the messages. In fact, it uses the paths contained in the deleting messages' list to calculate the number of times the relay nodes have met the final destination of messages. This technique enables the reduction of the overhead and increases the delivery probability of messages compared to FIFO strategy.

REFERENCES

[1] V. Cerf and al. 2001. *Interplanetary internet (IPN):* Architectural definition.

[2] A. Vahdat and D. Becker, (2000) *Epidemic routing for partially-connected ad hoc networks*, Tech. Rep.

[3] T. Spyropoulos, K. Psounis, and C. S. Raghavendra, (2005). *Spray and wait: Efficient routing in intermittently connected mobile networks*," in Proc.ACM SIGCOMM Workshop on Delay Tolerant Networking (WDTN).2005

[4] Phearin Sok, Keecheon Kim, (2013) *Distance-based PRoPHET Routing Protocol in Disruption Tolerant Network.*

[5] Ari Keränen, Jörg Ott and Teemu Kärkkäinen (2009) *The ONE Simulator for DTN Protocol Evaluation.*

[6] A. Lindgren and A. Doria ,(2012) *Probabilistic Routing Protocol for Intermittently Connected Networks.*

[7] Ashraf, Al-Fagih, Hossam and S.Hassanein; (2012) *Routing Schemes for Delay-Tolerant Networks* : An Applications Perspective Technical Report .

[8] J. Crowcroft, E. Yoneki, P. Hui, and T. Henderson,(2008) *Promoting tolerance for delay tolerant network research*, SIGCOMM.

11

The UWB Solution for Multimedia Traffic in Wireless Sensor Networks

A.A.BOUDHIR, M. BOUHORMA, M. BEN AHMED and ELBRAK SAID

LIST Laboratory, Faculty Of Sciences and Techniques, Tangier Morocco
CBAR, Al Yamamah university, Kingdom of Saudi Arabia

hakim.anouar@ieee.org , bouhorma@gmail.com, med.benahmed@gmail.com, elbraks@gmail.com

ABSTRACT

Several researches are focused on the QoS (Quality of Service) and Energy consumption in wireless Multimedia Sensor Networks. Those research projects invest in theory and practice in order to extend the spectrum of use of norms, standards and technologies which are emerged in wireless communications. The performance of these technologies is strongly related to domains of use and limitations of their characteristics. In this paper, we give a comparison of ZigBee technology, most widely used in sensor networks, and UWB (Ultra Wide Band) which presents itself as competitor that present in these work better results for audiovisual applications with medium-range and high throughput.

KEYWORDS

WMSNs, UWB, ZigBee, NS2

1. Introduction

The establishment of wireless networks in various domains knew an amazing success. In front of this progress several researches are followed to enlarge the range of its use. Networks of wireless sensors are a particular Ad hoc network, integrated with an active applications allowing control, surveillance and help to decision.

The introduction of wireless communication is dramatically changing our lives. The ability to communicate anytime anywhere increases our quality of lives and improves our business productivity. The recent technological developments that allow us to execute bandwidth-hungry multimedia applications over the wireless media add new dimensions to our ability to Communicate. Various technologies appeared in sensors networks and assure the communication differently, this difference comes especially in the given quality of service and solutions given to constraints.

In this document we study the quality of service of technologies Zigbee and UWB (Ultra Wide Band) as well as the consumption of energy for a multimedia flux using the simulator NS2.This to solicit the adapted technology to the transmission of Multimedia flux in WMSN(Wireless Multimedia Sensor Network).

2. RELATED WORKS

Several papers focused their works on WMNs treating an enormous topics related to energy efficiency, QoS, routing protocols on MAC and physical layers. Karapistoli *et al* [1] identify the cross-layer dependencies between the specified physical layer and the higher layers of the communication. In [2] authors raise a ranging method of localization technique in WSN based on ultra-wideband (UWB) communication technology. Melodia *et al* [3], present a cross-layer communication architecture based on the time-hopping impulse radio ultra wide band technology to deliver QoS to heterogeneous applications in WMSNs, by leveraging and controlling interactions among different layers of the protocol stack according to applications requirements. Berthe *et al* [4], propose a WSN simulation architecture based on the IR-UWB technique. At the PHY layer, they take into account the pulse collision by dealing with the pulse propagation delay. They also modelled MAC protocols specific to IRUWB,for WSN applications and propose a generic and reusable sensor and sensing channel model. Most of the WSN application performances can be evaluated thanks to this simulation architecture.

3. WIRELESS SENSOR NETWORKS

Wireless sensor networks (WSNs) contain hundreds or thousands of sensor nodes equipped with sensing, computing and communication abilities. Each node has the ability to sense elements of its environment, perform simple computations, and communicate among its peers or directly to an external base station or sink (figure.1). The node or mote that needs to operate for a long time on a tiny battery is composed of a processor, a memory, a transmitter/ receiver radio, an embedded system composed of a unit of sensing and a battery (figure2). This component can be in sleep mode or listen only to the traffic. The unit of transmission is the unit which uses most energy compared to others units constituting a sensor [05].

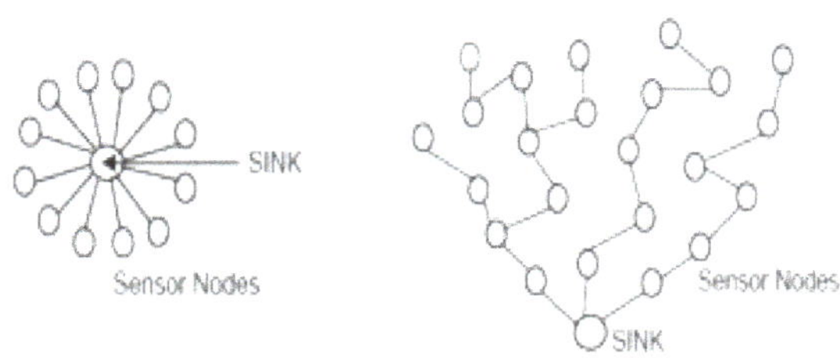

Figure.1: Architecture of Wireless Sensor Network

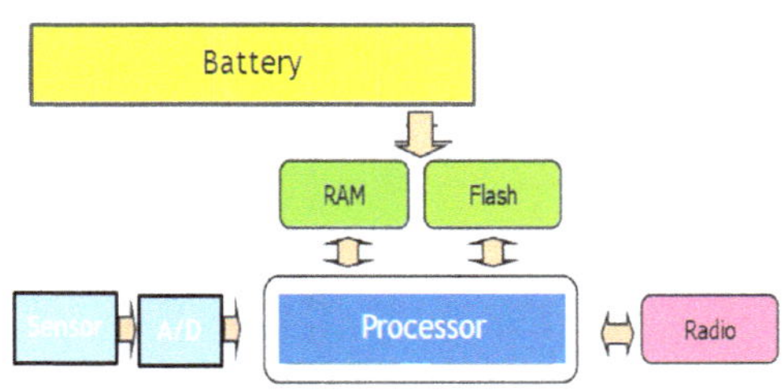

Figure.2: Architecture of Node

4. MULTIMEDIA WIRELESS SENSOR NETWORKS

The technology development in electronics contributed to the availability of miniaturized materials with low cost such as CMOS cameras and microphones which helped more the development of wireless multimedia sensor networks (WMSN), devices that are able to retrieve multimedia content such as the ubiquitous audio and video, still images, and data from environmental sensors.

Wireless multimedia sensor networks will not only reinforce the networks of sensors such as monitoring, home automation and environmental monitoring, but will also enable several new applications like monitoring networks multimedia, storage of potentially activities, control systems traffic, medical surveillance environmental monitoring, location services, industrial process control…

5. THE QUALITY OF SERVICE

In telecommunication networks, the goal of QoS is to reach a better behavior of communication for the content which must be properly routed, and network resources are used optimally [06]. Generally, researches on QoS in wireless networks in several key areas; models of QoS differentiation at the MAC layer (Medium Access Control) protocols for signaling and routing with QoS. The need of QoS can be specified into measurable parameters mentioned in (3), (4) and (5):

-End to End Delay:

$$EED = \frac{\text{Time spent to deliver packets}}{\sum \text{received packets}} \quad (3)$$

-Bandwidth:

$$BW = packets\ size\ x \frac{\sum \text{Received packets}}{\text{End Time Simulation}} \quad (4)$$

-Packet delivery ratio:

$$PDR\ (\%) = 100\ x \frac{\sum \text{Received packets}}{\sum \text{sent packets}} \quad (5)$$

6. TECHNOLOGIES EMERGED IN WIRELESS SENSOR COMMUNICATION

Many technologies are allowed to wireless transmission of information. Each represents a different use, according to its characteristics (transmission speed, maximum flow, Cost of infrastructure cost of equipment connected Security, Flexibility of installation and use, power consumption and autonomy …).

6.1. ZigBee Technology

ZigBee is a specification for a suite of high level communication protocols using small, low-power digital radios based on the IEEE 802.15.4 standard for wireless personal area networks (WPANs), such as wireless headphones connecting with cell phones via short-range radio. The technology defined by the ZigBee specification is intended to be simpler and less expensive than other WPANs, such as Bluetooth. ZigBee is targeted at radio-frequency (RF) applications such as industrial control and monitoring, wireless sensor networks, asset and inventory tracking, intelligent agriculture, and security would benefit from such a network topology that require a low data rate, long battery life, and secure networking.

ZigBee builds upon the physical layer and medium access control defined in IEEE standard 802.15.4 for low-rate WPAN's. The specification goes on to complete the standard by adding four main components: network layer, application layer, ZigBee device objects (ZDO's) and manufacturer-defined application objects which allow for customization and favor total integration.

ZigBee operates in the industrial, scientific and medical (ISM) radio bands; 868 MHz in Europe, 915 MHz in the USA and Australia, and 2.4 GHz in most jurisdictions worldwide. (figure.3)[07].

6.2. UWB Technology

Ultra Wide Band (UWB) technology based on sending pulses of energy low power over a wide frequency band is able to communicate wirelessly as an indoor short-range high-speed communication. One of the most exciting characteristics of UWB is that its bandwidth is over 110 Mbps (up to 480 Mbps) which can satisfy most of the multimedia applications, especially in wireless sensor networks, such as audio and video delivery in home networking and it can also act as a wireless cable replacement of high speed serial bus such as USB 2.0 and IEEE 1394. UWB works via chip-based radios that modulate signals across the entire available ultra wideband spectrum, which in the US is from 3.1 to 10.6 GHz (figure.4) [09]. In [08] several standards are mentioned giving the differences among four technologies. Each one is based on an IEEE standard.

Obviously, UWB and Wi-Fi provide a higher data rate, while Bluetooth and ZigBee give a lower one. In general, the Bluetooth, UWB, and ZigBee are intended for WPAN communication (about 10m), while Wi-Fi is oriented to WLAN (about 100m). However, ZigBee can also reach 100m in some applications.

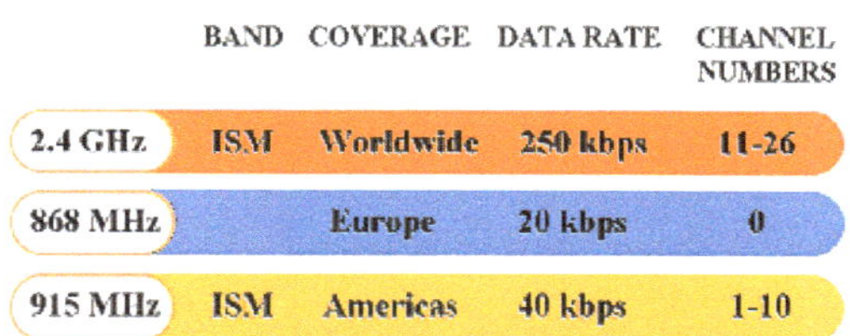

Figure.3: The three frequency band for IEEE 802.15.4 standard

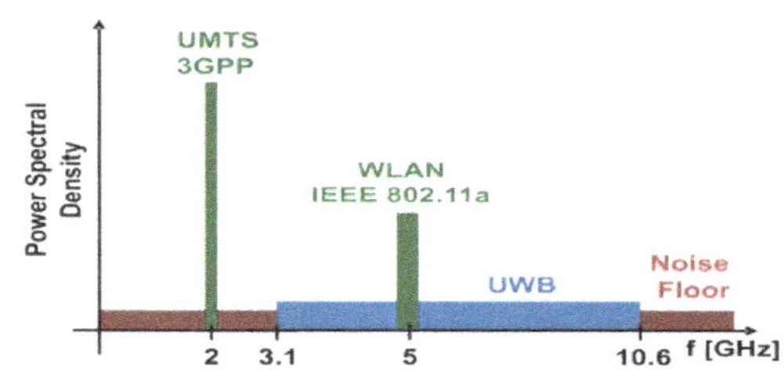

Figure.4: Frequency Spectrum in UWB technology

7. SIMULATION AND RESULTS

7.1. Environment Of Simulation

The simulation tool used is the NS2 simulator dedicated to wireless networks and considered a crucial asset search.

The version of ns2-allinone-2.29 [10]used, incorporate into the architecture of the MAC layer (mac.cc / mac.h) and physical (phy.cc / phy.h) modules and standard IEEE.802.15.4 supporting radio pulses compliant to IEEE 802.15.3 UWB which adds to its MAC layer modules DCC-MAC layer (mac-ifcontrol*. (cc, h)) and physical layer (interference-phy*. (cc, h)) by implementing the NOAH protocol that allows direct communications (unlike AODV, DSR, ...) between wireless nodes, or between base stations and mobile nodes. It can simulate scenarios where multi-hop routing is undesirable.

**Mac IFcontrol* [10]: Defines the MAC layer for UWB, functions of transmission, queue management, control packets and listening mode etc.

**Interference-phy* [10]: defines possible states (reception, transmission, listen or hang) and manages the time to listen and reception etc.

7.2. PARAMETERS OF SIMULATIONS

In order to evaluate the quality of service, the simulations treat a comparative metric subject to two different multicast protocols AODV [11], [12] and DSR [11],[12] for ZigBee technology and protocol NOAH for UWB parameters as mentioned below. The figure 5 list an example of architecture adopted, using node as wireless multimedia sensors and a collector of data "C" (node0 in NAM Visualisator). The simulation results are drawn from the files "tr" generated and analyzed by file "awk".
The table 2 summarizes the simulation parameters used for the ZigBee and UWB technologies:

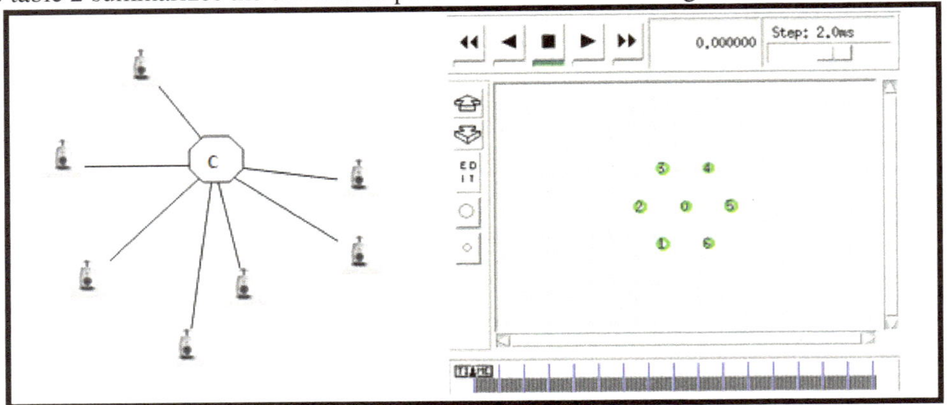

fig.5: Sample of architecture adopted in simulations

TABLE 2
Parameters used in ZigBee and UWB simulations

Technology	ZigBee	UWB
Protocol	AODV / DSR	NOAH
Mac /phy	802_15_4	802_15_3
Chanel	Wireless Channel	InterferencePhy
Propagation	TwoRayGround	PropTarokh
Topology	20*20	20*20
Traffic	UDP/FTP	UDP/ FTP
Number of nodes	7 – 25 - 101	7 - 25 - 101
RtPower	0.00075 w	0.00075 w
TxPower	0.00175 w	0.00175 w
Initial Energy	1000 j	1000 j
Sleep Energy	0.00005j	0.00005j

7.3. RESUALTS AND DISCUSSION

The following figures illustrate a comparison based on the results of previous simulations show the benefit of UWB in the delivery of a high rate of packets (fig7 and fig8) compared to ZigBee (fig6) with a wide bandwidth (fig9) and a delay (fig10) start to finish while consuming a minimum of Energy (fig11)for less dense networks:

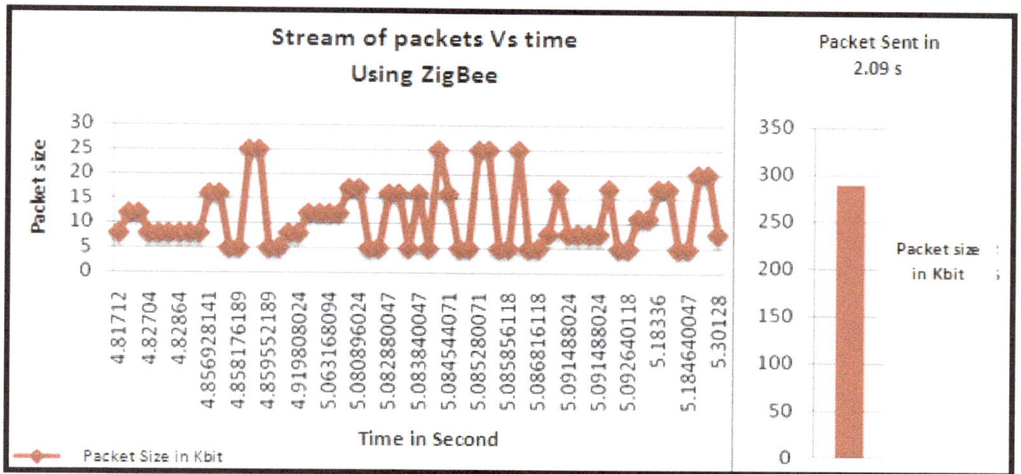

fig.6 : Stream Packets Vs Time Using ZigBee

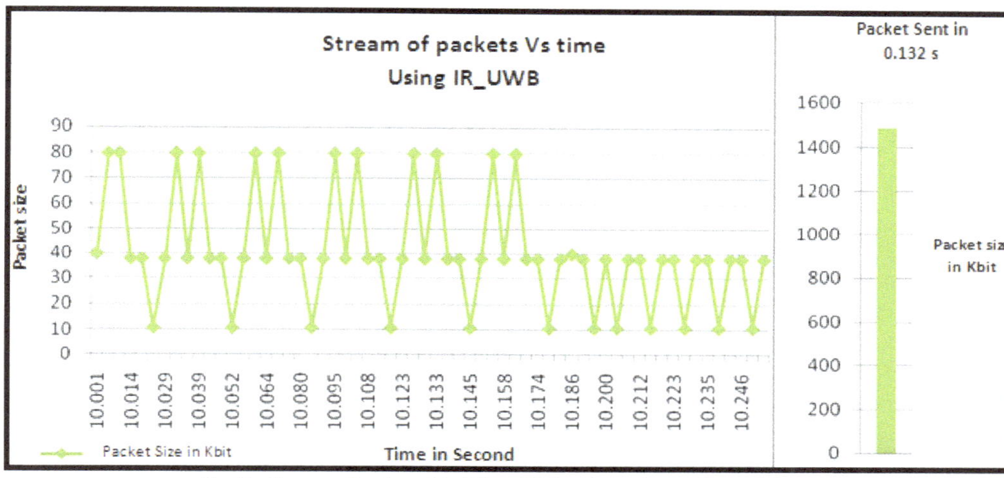

fig.7 : Stream Packets Vs Time Using impulse radion UWB

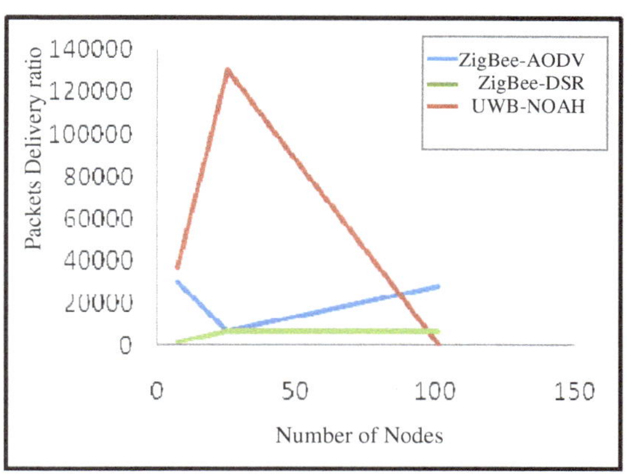

Fig.8: Packets Delivery Ratio

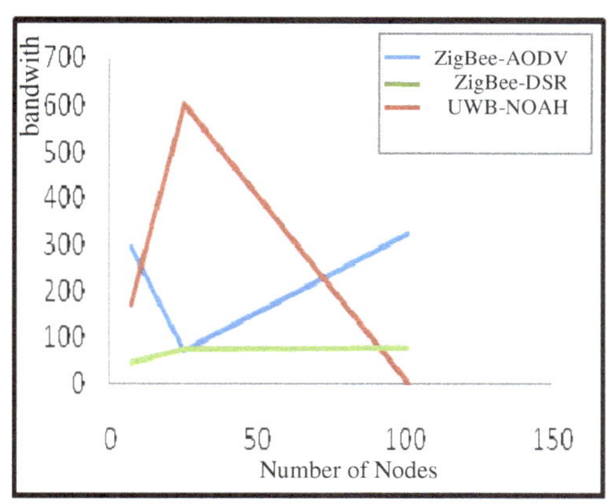

Fig.9: Bandwidth

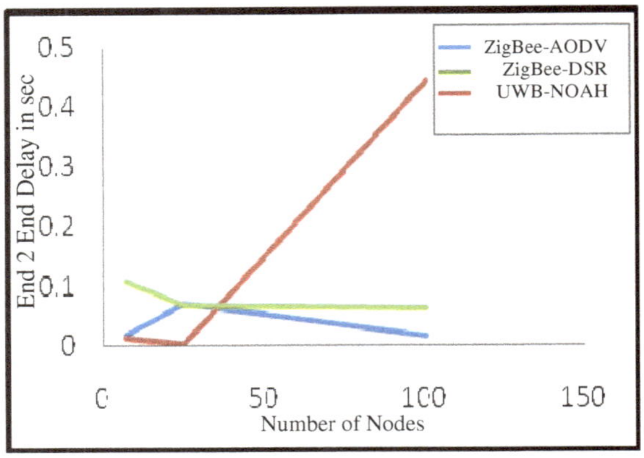

Fig.10: End to End Delay

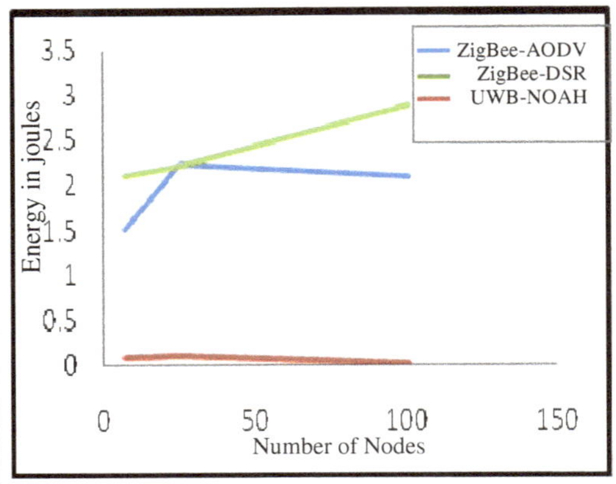

Fig.11: Energy Consumption

With the simulation parameters mentioned, the UWB technology offered a considerable quality of service in term of end to end delay, packet delivery ratio, bandwith and energy for a sensor network with less density of network when sending media streams. Indeed, this technology has gives a good results compared to those of the ZigBee technology especially for a network average of 25 nodes which promotes greater use of these sensors for the transfer of multimedia data.

8. CONCLUSION

This work has explored a survey on UWB and ZigBee standars in order to compare their performance in wireless multimedia sensor network context. We have presented in this article a quality of service of standards mentionned above, using the popular metrics as a network performance tools. In particular, those metrics, already presented, were evaluated to study the quality of communication of multimedia data under the network simulator NS2.

Furthermore, we found that the performance tests conducted on the consumption of energy, packet delivery ratio, bandwidth and end-to-end delay, have shown that the UWB technology responds well performance criteria desired. Indeed, the ultra wide band (UWB) technology has the potential to enable low-power consumption, high data rate communications within tens of meters, characteristics that make it an ideal choice for WMSNs.

REFERENCES

[1] Karapistoli, Eirini; Gragopoulos, Ioannis; Tsetsinas, Ioannis; Pavlidou, Fotini-Niovi "UWB Technology to Enhance the Performance of Wireless Multimedia Sensor Networks " in Computers and Communications, 2007. ISCC 2007. 12th IEEE Symposium on

[2] Chunsen Xu, Yanjuan Zhao, Yan Zhang in ,"Localization Technology in Wireless Sensor Networks Based on UWB" . International Conference on Wireless Networks and Information Systems (2009)

[3] T Melodia, I F Akyildiz "Cross-Layer Quality of Service Support for UWB Wireless Multimedia Sensor Networks " in IEEE INFOCOM The 27th Conference on Computer Communications (2008)

[4] Berthe, A.; Lecointre, A.; Dragomirescu, D.; Plana, R. "Simulation Platform for Wireless Sensor Networks Based on Impulse Radio Ultra Wide Band Networks" ICN '09. Eighth International Conference on Digital Object Identifier: 10.1109/ICN.2009.48 (2009).

[5] Yacine CHALLAL "Réseaux de Capteurs Sans Fil" –University of Technology in compiegne, France.

[6] Wendi Rabiner Heinzelman, Anantha Chandrakasan, and Hari "Energy-Efficient Communication Protocol for Wireless Microsensor Networks" Proc. of the 33rd Hawaii International Conference on System Sciences – 2000.

[7] Houarrii MAOUCHI " Routage avec Qualité de Service dans AODV ", thesis presented in university Mouloud MAMMERI of tizi-ouzou.

[8] Jon Adams "Intelligent Systems Wireless Networking", Motorola Wireless and Broadband Systems. Available: http://archives.sensorsmag.com/articles/0603/14/.

[9] Jin-Shyan Lee, Yu-Wei Su, and Chung-Chou Shen "A Comparative Study of Wireless Protocols: Bluetooth, UWB, ZigBee, and Wi-Fi" Conf. of the IEEE Industrial Electronics Society (IECON) Nov. 5-8, 2007, Taipei, Taiwan.

[10] Niels Hadaschik,"Techniques for UWB-OFDM" Available: http://www.iss.rwth-aachen.de/Projekte/theo.

[11] Jean-Yves Le Boudec,Ruben Merz,Bozidar Radunovic,Jörg Widmer "NS-2 (UWB) MAC and PHY simulator " Available :http://icawww1.epfl.ch/uwb/ns-2/index.html

[12] Rachid Ennaji "Routing in Wireless Sensor Networks" 978-1-4244-3757-3/09/$25.00©2009IEEE-School of Science and Engineering- Al Akhawayn University in Ifrane, Morocco

ANALYSIS OF SECURITY THREATS IN WIRELESS SENSOR NETWORK

Sahabul Alam and Debashis De

Department of Computer Science and Engineering,West Bengal University of Technology, Kolkata,India

ABSTRACT

Wireless Sensor Network(WSN) is an emerging technology and explored field of researchers worldwide in the past few years, so does the need for effective security mechanisms. The sensing technology combined with processing power and wireless communication makes it lucrative for being exploited in abundance in future. The inclusion of wireless communication technology also incurs various types of security threats due to unattended installation of sensor nodes as sensor networks may interact with sensitive data and /or operate in hostile unattended environments. These security concerns be addressed from the beginning of the system design. The intent of this paper is to investigate the security related issues in wireless sensor networks. In this paper we have explored general security threats in wireless sensor network with extensive study.

KEYWORDS

Wireless Sensor Networks,Security,Threats,Attacks

1. INTRODUCTION

Wireless sensor networks are network of thousand of sensor nodes. Sensor nodes are small in size, less memory space, cheaper in price with restricted energy source and limited processing capability. WSNs are rapidly gaining popularity due to low cost solutions to a variety of real world challenges [1].The basic idea of sensor network is to disperse tiny sensing devices, which are capable of sensing some changes of incidents / parameters and communicating with other devices over a specific geographic area for some specific purposes like surveillance, environmental monitoring, target tracking etc. Sensor can monitor pressure,humidity, temperature, vehicular movement, lightning conditions, mechanical stress levels on attached objects and other properties [2].Due to the lack of data storage and power sensor networks introduce severe resource constraints. These are the obstacles to the implementation of traditional computer security techniques in a WSN. Security defenses harder in WSN due to the unreliable communication channel and unattended operation. As a result these networks require some unique security policies.Cryptography,steganography and other basics of network security and their applicability can be used to address the critical security issues in WSN. Many researchers have begun to address of maximizing the processing capabilities and energy saving of sensor nodes with securing them against attackers.There are different types of attacks designed to exploit the unreliable communication channels and unattended operation of WSNs. Physical attacks to sensors play an important role in the operation of WSNs due to the inherent unattended feature. We explore various types of attacks and threats against WSN.

2. FUNDAMENTAL SECURITY SCHEMES IN WIRELESS SENSOR NETWORK

Security encompasses the characteristics of authentication, privacy,integrity,anti-playback and nonrepudiation. The more the risk of secure transmission of information over the network has increased as the more dependency on the information provided by the networks has been increased.Several cryptographic, steganographic and other techniques are used which are well known for the secure transmission of various types of information over networks [3].In this section, we discuss the network security basic and how the techniques are meant for wireless sensor networks.

2.1 Cryptography

The encryption-decryption techniques devised for the traditional wired networks are not feasible to be applied directly for wireless sensor networks. WSNs consist of tiny sensors which really suffer from the lack of battery power,processing and memory. Any encryption scheme applying on WSNs require transmission of extra bits,hence extra processing,memory and battery power which are very important resources for the sensor's longevity. Applying the security mechanisms such as encryption could also increase delay, jitter and packet loss in wireless sensor networks [3]. There are some key questions arise when applying encryption schemes to WSNs like, how the keys are generated or disseminated. There is an important issue how the keys could be modified time to time for encryption as there is minimal (or no) interaction for the sensors. There are other many issues how keys are revoked, assigned to a new sensor added to the network or renewed for ensuring robust security for the network. There could not be an efficient solution for adopting of pre-loaded keys or embedded keys.

2.2 Steganography

Steganography aims at hiding the existence of the message while cryptography aims at hiding the content of a message. In steganography a message embeds into the multimedia data(image,sound,video etc.).Steganography modifies the carrier in a way that is not perceptible and looks just like ordinary. It is very useful when we want to distribute secret data publicly and in the case that we want to send a secret data without sender information as it hides the existence of the covert channel. Securing wireless sensor networks is not directly related to steganography and processing multimedia data (like audio,video). With the inadequate resources of the sensors for securing wireless sensor networks is difficult and an open research issue.

2.3 Physical Layer Secure Access

Frequency hopping provides physical layer secure access in WSN. A dynamic combination of the parameters like hopping set(available frequencies for hopping), hopping pattern (the sequence in which the frequencies from the available hopping set is used) and dwell time(time interval per hop) could be used with a little expense of memory, processing and energy resources. In physical layer secure access the efficient design is required so that the hopping sequence is modified in less time than to discover it and for employing this both the sender and receiver should maintain a synchronized clock. A scheme as proposed in [4] could also be utilized which introduces secure physical layer employing the singular vectors with the channel synthesized modulation.

3. THREAT MODEL

It is usually assumed that an attacker may know the security mechanisms that are deployed in a sensor network. Attackers may be able to compromise a node or even physically capture a node. Most WSN nodes are viewed as non-tamper resistant due to the high cost of deploying tamper-resistant sensor nodes. The attacker is capable of stealing the key materials contained within the compromised node. Base stations are regarded as trustworthy in WSNs. Most researchers focus on secure routing between sensors and the between base stations.

Attacks in sensor networks can be classified into the following categories

1. Outsider Vs. insider attacks: Outsider attacks are attacks from nodes which do not belong to a WSN. Insider attacks occur when legitimate nodes of a WSN behave in unintended or unauthorized ways.

2. Passive Vs. active attacks: Passive attacks include eavesdropping on or monitoring packets exchanged with in a WSN. Active attacks involve some modifications of the data stream or the creation of a false stream.

3. Mote-class Vs. Laptop-class attacks: An adversary attacks a WSN by using a few nodes with similar capabilities to the network nodes in mote-class attacks. Mote class attackers can jam the radio link in it's immediate vicinity. In laptop class attacks an adversary can use more powerful devices (e.g a laptop) to attack a WSN. These devices have greater transmission range, energy reserves and processing power than the network nodes. A laptop class attacker might be able to eavesdrop on an entire network.

4. THREAT ATTACKS IN WIRELESS SENSOR NETWORK

Why is security necessary in WSN? Due to the broadcast nature of the transmission medium wireless sensor networks are vulnerable. There are another reason of vulnerability of WSNs are nodes are often placed in a hostile or dangerous environment and they are not physically safe. Most of the threats and attacks against security in wireless sensor networks are almost similar to their wired counterparts while some are exacerbated with the inclusion of wireless connectivity. WSNs are usually more vulnerable to various security threats because the unguided transmission medium is more susceptible to security attacks, but also through traffic analysis, privacy violation, physical attacks and so on.

Attacks on WSNs can be classified from two different levels of views

1. Attack against security mechanisms
2. Attack against basic mechanisms (like routing mechanisms)

In many applications the data obtained by the sensing nodes need to be authentic [5]. A false or malicious node could intercept private information in the absence of proper security or could send false messages to nodes in the network. Different possible attacks can be categorized as follows

4.1. Denial of Service Attacks

In WSN, Denial of Service (DOS) is produced by the unintentional failure of nodes or malicious action. In DOS attack the adversary attempts to subvert, disrupt or destroy a network. DOS attack diminishes a network capability to provide a service for any event. The simplest DOS attack tries to exhaust the resources available to the victim node, by sending extra unnecessary packets and thus prevents legitimate network users from accessing services or resources to which they are entitled [6].

4.1.1 Jamming

Jamming is a DOS attack at physical layer. Jamming interferes with the radio frequencies that a network's nodes are using [7]. A jamming source may either be powerful enough to disrupt the entire network or less powerful and only able to disrupt a smaller portion of the network. It creates radio interference and resource exhaustion.To defend against jamming involve variations of spread-spectrum communication such as frequency hopping and code spreading.Security class of this attack is modification.

4.1.2 Tampering

Another DOS attack in physical layer is tampering. By physical access an attacker can extract sensitive information such as cryptographic keys or other data on the node. A compromised node creates, which the attacker controls by altering or replacing node. Vulnerability of this attack is logical.One defence to this attack involves tamper-proofing the node's physical package.

4.1.3 Collisions

Collision is a DOS attack in the data link layer. When two nodes attempt to transmit on the same frequency simultaneously a collision occurs. A change will likely to occur in the data portion when packets collide and causing a checksum mismatch at the receiving end. The packet will then be discarded as invalid. An adversary may strategically cause collisions in specific packets such as ACK control messages. Error-correcting codes use to defend against collisions.

4.1.4 Exhaustion

It is another type of DOS attack in link layer. An attacker can use repeated collisions to cause resource exhaustion. For example, a native link-layer implementation may continuously attempt to retransmit the corrupted packets. The energy reserves of the transmitting node unless these hopeless retransmissions are discovered or prevented. Applying rate limits to the MAC admission control is a possible solution of exhaustion.

4.1.5 Unfairness

Unfairness is a weak of a DOS attack in link layer. An attacker, may cause unfairness in a network by using the above link- layer attacks. Instead of preventing access to a service outright, an attacker can degrade it in order to gain advantage such as causing other nodes in a real time MAC protocol to miss their transmission deadline.

4.1.6 Flooding

Flooding is a DOS attack in transport layer. A protocol becomes vulnerable to memory exhaustion through flooding when it maintains at either end of a connection. An attacker may repeatedly make new connection requests until the resources required by each connection are exhausted or reach a maximum limit. In either case, further legitimate requests will be ignored. Disrupt communication is one of purpose of this attack. It creates resource exhaustion and reduces availability.

4.1.7 Desynchronization

Disruption of an existing connection is desynchronization. For example, an attacker may repeatedly spoof messages to an end host, causing that host to request the retransmission of missed frames. With proper timing, an attacker may degrade or even prevent the ability of the end hosts to successfully exchange data. A possible solution to this type of attack is to require authentication of all packets communicated between hosts. The authentication method would be secure as an attacker will be unable to send the spoofed messages to the end hosts.Vulnerability of this attack is logical.

4.1.8 Data Integrity Attack

Data integrity attacks are caused by changing the data contained within the packets or injecting false node while the data travelling among the nodes in WSN. The attacker node must have more processing, memory and energy than the sensor nodes. The goals of this attack are to falsify routing data in order to disrupt the sensor network's normal operation and also falsifies sensor data by doing so compromise the victim's research. Digital signatures and asymmetric key systems are used to defend against this attack. This requires a lot of additional overhead and is difficult to adapt in WSN.

4.2 The Sybil Attack

The sensors in a WSN might need to work together to accomplish a task in many cases, hence they can use distribution of subtasks and redundancy of information. In such case a single node duplicates itself and presented in the multiple locations. The Sybil attack targets fault tolerant schemes such as multipath routing, distribute storage and topology maintenance. Sybil attack tries to degrade the integrity of data security and resource utilization that the distributed algorithm attempts to achieve. Any peer to peer network especially wireless adhoc network is vulnerable to sybil attack. Encryption and authentication techniques can prevent an outsider to launch a sybil attack on the sensor network. If a compromised node pretends to be two of the three nodes the algorithms used may conclude that redundancy has been achieved while in reality it has not. Public key cryptography can prevent an insider attack but it is too expensive to be used in the resource constrained sensor network. Figure 1 shows sybil attack.

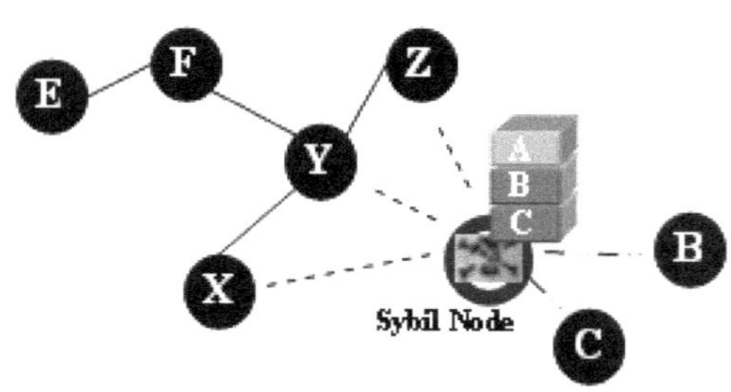

Figure 1. Sybil Attack

4.3 Blackhole Attack

A malicious node acts as a blackhole[8] in the range of the sink attracts the entire traffic to be routed through it by advertising itself as the shortest route. The adversary drops packets coming from specific sources in the network. Once the malicious device is in between the communicating nodes (for example, sink and sensor node), it is able to do anything with the packets passing between them. This attack can also affect the nodes those are considerably far from the base stations. It creates high rate of packet loss,network partition.It decreases the throughput of a subset of nodes.The network architecture of this attack is traditional wireless sensor network. Figure 2 shows the conceptual view of a blackhole attack.

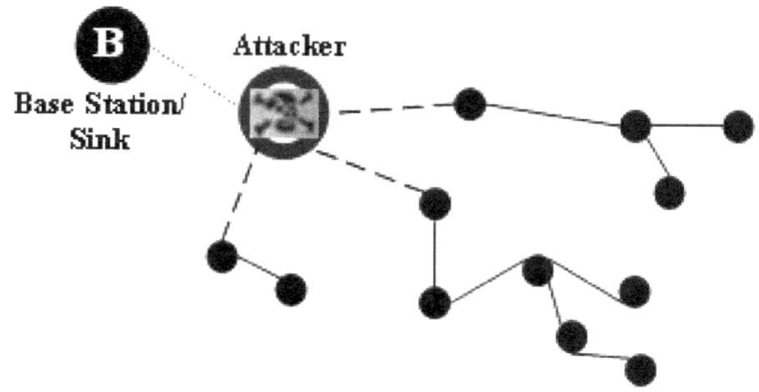

Figure 2. Blackhole Attack

4.4 HELLO Flood Attack

HELLO Flood is a novel attack against sensor networks. It is introduced in [9].This attack uses HELLO packets as a weapon to convince the sensors in WSN. Many protocols require nodes to broadcast HELLO packets to announce themselves to their neighbours and a receiving node may assume that it is within normal radio range of the sender. This assumption may be false because a laptop class attacker broadcasts routing or other information with large enough transmission power could convince every node in the network that the adversary is it's neighbour. For example, an adversary advertises a very high quality route to the base station to every node in the network could cause a large number of nodes to attempt to use this route, but those nodes sufficiently far away from the adversary would be sending packets into oblivion. The network is left in a state of confusion.This attack wastes node's energy due to laptop class attacker and creates routing delay. The network architecture of this attack is traditional wireless sensor network.

4.5 Traffic Analysis Attack

WSNs consist of many low-power sensors communicating with a few powerful and robust base stations. Data gathered by the individual nodes routed to the base station. Often, for an adversary to effectively render the network useless, the attacker can simply disable the base station. Even when encrypted messages are transferred, it still leaves a high possibility analysis of the communication patterns. Sensor activities can potentially reveal enough information to enable an adversary to cause malicious harm to the sensor network. Table 1shows layer based attacks and possible security mechanisms in wireless sensor networks.

Layer	Attacks	Security Approach Defenses
Physical layer	Jamming	Lower duty cycle, Priority messages, Spread-spectrum techniques
	Tampering	Tamper proofing, Hiding
Data Link layer	Collision	Error-correcting code
	Exhaustion	Rate limitation

	Unfairness	Small frames
Network layer	Spoofed, altered or replayed routing information	Authentication, Monitoring
	Selective forwarding	Probing, Redundancy
	Sink hole	Monitoring, Redundancy, Authentication
	Sybil	Probing, Authentication
	Worm holes	Authentication, Packet leashes by using geographic and temporal information
	Hello flood	Verify the bidirectional link, Authentication
	Acknowledgment spoofing	Authentication
Transport layer	Flooding	Client puzzles
	Desynchronization	Authentication
Application layer	Attacks on reliability and Clone attack: Clock skewing, Selective message forwarding, Data aggregation distortion	Unique pair wise keys and cryptographic approach. Authentication can be used to protect any data integrity Encryption is an effective approach for data confidentiality protection

Table 1. Layer based attacks and possible security mechanisms in wireless sensor network

4.6 Node Replication Attack

It is quite simple attack. An attacker copies the node ID of an existing sensor node and adds to an existing sensor network. With node replication approach a sensor network's performance can severely disrupt. Packets can be corrupted or even misrouted. As a result this can form a disconnect network, false sensor reading etc. An attacker can copy cryptographic keys to the replicated sensors by physical accessing to the entire network. The attacker could easily manipulate a specific segment of the network by inserting the replicated nodes at specific network points, perhaps by disconnecting it altogether.

4.7 Attacks Against Privacy

Sensor network technology provides automatic data collection capabilities through efficient deployment of tiny sensor devices. These technology exhibits significant potential for abuse. As sensor network provides increased data collection capabilities, so privacy problems arise. The main privacy problem is not that sensor network enable the collection of information. Direct site surveillance uses to collect much information from sensor networks. Sensor networks suffer privacy problems because they make large volume of information easily available through remote access. Adversaries can gather information at low-risk in anonymous manner because they need not be physically present to maintain surveillance.

4.8 Physical Attacks

Sensors networks typically operate in hostile outdoor environments. The sensor networks are highly susceptible to physical attacks, i.e. threats due to physical node destructions as sensors are small in size, deployed with the unattended environment. Physical attack destroys sensors permanently, so there are looses of cryptographic secrets, tamper with the associated circuitry, modify or replace sensors with malicious sensors under control of the attacker.

4.9 Wormhole Attack

Wormhole attack is a significant attack in WSN. This attack occurs at the initial phase when the sensors start to discover the neighbouring information [10].In this attack there is no need compromising a sensor in the network. In wormhole attack, the attacker records the packets (or bits) at one location in the network and tunnels those to another location. The tunnelling or retransmitting of bits could be done selectively. The simplest case of this attack is to have a malicious node forwarding data between two legitimate nodes. Wormholes convince distance nodes that they are neighbours, leading to quick exhaustion of their energy resources. An adversary situated close to a base station may be able to disrupt routing by creating a well-placed wormhole.

4.10 Spoofed, Altered or Replayed Routing Information

This is the most common direct attack against a routing protocol. This attack mainly targets the routing information exchanged between the nodes. In order to disrupt traffic in the network, an attacker may spoof, alter or reply routing information. These disruptions include the creation of routing loops, extending and shortening source routes, attracting or repelling network traffic from select nodes, partitioning the network, generating fake error messages and increasing end-to-end latency. Authentication is the standard solution for these attack i.e routers will only accept routing information from valid routers.

4.11 Selective Forwarding Attack

Multi-hop mode of communication is commonly preferred in WSN data gathering protocols. An assumption made in multihop networks is that all nodes in the network will accurately forward received messages. Selective forwarding attack is a situation when certain nodes do not forward many of the messages they receive. In this attack, malicious nodes may refuse to forward certain messages and simply drop them, ensuring that they are not propagated any further. When a malicious node behaves like a black hole and refuses to forward every packet she sees is a simple form of selective forwarding attack. This attack can be detected if packet sequence numbers are checked properly and continuously in a conjunction free network. Data packet sequence number in packet header can reduce this type attack. Vulnerability of this attack is logical.We have done comparative analysis of different threats in wireless sensor networks in table 2.

Attacks	Principle[11]	Disadvantage	Network architecture	Vulnerability[14]	Security class[15]
Denial of Service (DoS) attacks	All Purposes	Eliminates the network's capacity	Traditional wireless sensor network	Logical	Interruption, interception, modification, fabrication
Wormholes	Unfairness, to be authenticated, to be authorized	Difficult to check routing information	Traditional wireless sensor network	Logical	Fabrication, interception
Spoofed, altered or replayed routing information	Unfairness	Generates false error messages, increase end-to-end latency, extend or shorten source routes	Large, distributed and traditional wireless sensor network	Logical	Fabrication, modification
Link layer Jamming	Disrupt communication	Exhausting node's resources, confusion and packets collision	Traditional wireless sensor network	Logical	Modification
Collision	Unfairness	Energy exhaustion, discarding packets, interefences, cost effective	Large and traditional wireless sensor network	Logical	Modification
Resource Exhaustion	Unfairness	Compromise availability, Resource exhaustion	Traditional wireless sensor network	Logical	Modification
Unfairness	Unfairness	Node's hungry to channel access, Efficiency and utility of service decrease	Traditional wireless sensor network	Logical	Modification

Acknowledge spoofing	Unfairness	False information to the neighboring nodes, packet loss,selective forwarding attack	Traditional and distributed wireless sensor network	Logical	Fabrication, modification
Sybil	Unfairness, To be authenticated, To be authorized	Launch threat to geographic routing protocol, Expensive in sensor network	Traditional wireless sensor network	Logical	Modification, Fabrication

Principle[11]: Passive eavesdrop,disrupt communication, unfairness, to be authorized, to be authenticated

Vulnerabilities[14]: Physical (hardware) ,logical
Security class[15]: Interruption, interception, modification, fabrication

Table 2. Comparative analysis of different threats in wireless sensor networks

4.12 Passive Information Gathering

With a powerful receiver and well designed antenna an intruder can easily pickoff the data stream. An attacker locates the nodes and destroy them by interception of the messages containing the physical location of sensor networks[11]. An adversary can observe the application specific content of messages including message Ids, timestamps and other fields besides the location of sensor nodes.

4.13 Node Capturing

A particular sensor might be captured. Node capturing may reveal it's information including disclosure of cryptographic keys and an adversary can get information from the captured node [12].Node capturing compromises the whole sensor network and prevents from any communication. It launches a variety of insider attacks, software vulnerabilities.

4.14 False or Malicious Node

Attacks against security in WSN are caused by the insertion of false information from compromised nodes within the network [12].An intruder might add a malicious node to the system that feeds false data or prevents the passage of true data. The most dangerous attack is insertion of malicious node. Injecting malicious nodes destroy the whole network.

4.15 Energy Drain Attack

WSN is battery powered and dynamically organized. It is difficult or impossible to replace/recharge sensor node batteries. Attackers may use compromised nodes to inject fabricated reports into the network or generate large amount of traffic in the network as limited amount of energy available in sensor node. Fabricated reports will cause false alarms that waste real world response efforts and drain the finite amount of energy in a battery powered network. The network architecture of this attack is traditional wireless sensor network. This attack is possible only if the intruder's node has enough energy to transmit packets at a constant rate. This attack destroys the sensor nodes in the network, degrades performance of the network and ultimately splits the network grid and by inserting a new sink node takes control of part of the sensor network. Fabricated reports should be dropped en-route as early as possible to minimize the damage caused by this attack.

4.16 Acknowledgement Spoofing

Mostly sensor network routing algorithms rely on implicit or explicit link layer acknowledgements. An adversary can spoof link layer acknowledgments for 'overhead' packets addressed to neighbouring nodes due to the inherent broadcast medium. Protocols that choose the next hop based on reliability issues is susceptible packets being lost when travel along such links. The goal includes convincing the sender that a weak link is strong or that a dead or disabled node is alive. Acknowledgment spoofing attacks can be prevented by proper authentication for communication and using good encryption techniques [13].

5. CONCLUSIONS

Most of the attacks in wireless sensor networks are caused by the insertion of false information. For defending the inclusion of false reports by compromised nodes is required detecting mechanism. Developing such detection mechanism is great research challenge. All of the previously mentioned security threats i.e the HELLO flood attack, wormhole attack, sinkhole attack, Sybil attack serves one common purpose that is to compromise the integrity of the network they attack. In the past focus has not been on the security of WSNs. Security has become a major issue for data confidentiality as the various threats are arising. In this paper we have tried to present most of the security threats in WSN with extensive study.

6. REFERENCES

[1] I.F.Akyildiz, W.Su, Y.Sankarasubramaniam and E.Cayirci , A survey on sensor networks, IEEE Communications Magazine, 40(8): 102-114, August 2002.

[2] Pathan, A-S.K,Islam,H.K.Sayeed,S.A.Ahmed,F.and Hong, C.S," A Framework for Providing E-Services to the Rural Areas using Wireless Ad Hoc and Sensor Networks,"to appear in IEEE ICNEWS 2006.

[3] Kumar,P,; Cho,S.;Lee, Y.D.;Lee, H.J.TriSec ," A secure data framework for wireless sensor networks using authenticated encryption,"Int.J.Marit.Inf.Commun.Sci(2010),129-135.

[4] Orihashi,M.,Nakagawa,Y.,Murakami,Y.,and Kobayashi,K.," Channel synthesized modulation employing singular vector for secured access on physical layer", IEEE GLBOECOM 2003,volume 3,1-5 December,2003,pp.1226-1230.

[5] Adrian Perrig,John Stankovic, and David Wagner,(2004)" Security in wireless sensor networks",Commun;ACM,4796):53-57.

[6] A.D.Wood and J.A.Stankovic,(2002)" Denial of Service in Sensor Networks,"Computer,vol.35,no.10,2002,pp.54-62.

[7] E.Shi and A.Perrig," Designing Secure Sensor Networks," Wireless Commun.Mag,vol.11,no.6,Dec. 2004,pp.38-43.

[8] Culpepper,B.J.and Tseng,H.C.," Sinkhole intrusion indicators in DSR MANETs",Proc.First International Conference on Broadband Networks,2004,pp.681-688.

[9] Kerlof,C. and Wagner,D.," Secure routing in wireless sensor networks:Attacks and countermeasures",Elsevier's Ad Hoc Network Journal, Special Issue on Sensor Netwok Applications and Protocols , September 2003,pp.293-315.

[10] Zaw Tun and Aung Htein Maw,(2008),"Wormhole Attack Detection in Wireless Sensor networks", Proceedings of World Academy of Science ,Engineering and Technology,Volume 36,December 2008,ISSN 2070-3740.

[11] Al-Sakib Khan Pathan et al.,(2006)" Security in Wireless sensor networks: Issues and Challenges", in feb. 20-22,2006,ICACT2006,ISN 89-5519-129-4,pp(1043-1048).

[12] C.Kar lof and D.WAGNER,(2003),"Secure routing in wireless sensor networks: Attacks and countermeasures",AdHoc Networks Journal,vol.1,no.2-3,pp.293-315,September.

[13] P.Mohanty,S.Panigrahi,N.Sharma and S.Satapathy," Security Issues in Wireless Sensor Network Data Gathering Protocols:A Survey",in Journal of Theoetical and Applied Information Technology(JATIT).

13

PEER TO PEER CACHE RESOLUTION MECHANISM FOR MOBILE AD HOC NETWORKS

Preetha Theresa Joy[1] and K. Poulose Jacob[2]

[1,2] Department of Computer Science, Cochin University of Science and Technology, Kochi, Kerala, India.

preetha@mec.ac.in, kpj@cusat.ac.in

ABSTRACT

In this paper we investigate the problem of cache resolution in a mobile peer to peer ad hoc network. In our vision cache resolution should satisfy the following requirements: (i) it should result in low message overhead and (ii) the information should be retrieved with minimum delay. In this paper, we show that these goals can be achieved by splitting the one hop neighbours in to two sets based on the transmission range. The proposed approach reduces the number of messages flooded in to the network to find the requested data. This scheme is fully distributed and comes at very low cost in terms of cache overhead. The experimental results gives a promising result based on the metrics of studies.

Keywords

Cache Resolution, MANET, Cooperative caching, Wireless Networks.

1. INTRODUCTION

Recent wireless communication growth and the continuous improvement of mobile terminal performance have enabled the use of these technologies in different fields and to consider new applications. There are two categories of wireless mobile networks: networks with infrastructure using access points to connect to a wide range of network and networks without infrastructure which do not presuppose the presence of a wired infrastructure. An ad hoc network is a wireless self configurable network made up of mobile free moving terminals interconnected by wireless connections. Ad hoc networks provide a way of communicating which can be quickly and easily deployed. This makes the ad hoc networks a good choice for applications in which various civil and military fields.

The goal of a computer network is to offer a way for users with different terminals to communicate in order to share data and access services provided by different servers. In mobile ad hoc networks, devices generally have limited energy reserves and processing capabilities. Bandwidth is also a scarce resource, limited by the nature of the wireless medium. In a data-management point of view, these restrictions introduce several issues that need to be addressed. Data transfers must be reduced and mechanisms must be deployed to confront the frequent disconnections and low bandwidth constraints. Therefore it is a challenging task to present the data efficiently by reducing the delay or waiting time to the end user.

Data caching is widely used in various domains to improve data access efficiency, by reducing the latency experienced by the end users. In wireless mobile network, holding frequently accessed data items in a mobile node's local storage can reduce network traffic, response time and server load. Caching in ad hoc networks is effective because a few resources are requested often by many users, or repeatedly by a specific user, which is known as the locality of reference. To have the full benefits of caching, the neighbor nodes can cooperate and serve each other's misses, thus further reducing the wireless traffic. This process is called cooperative caching. Since the mobile nodes can make use of the data stored in another node's cache the effective cache size is increased.

The implementation of a good cooperative caching technique essentially involves four major design considerations: cache placement and resolution, cache admission control, cache replacement and cache consistency maintenance [1]. The cache admission control module decides whether a received data item is cacheable or not, and a cache resolution module decides how to fetch the data from the neighboring nodes when there is a local miss. Since the network node has limited memory, it can cache multiple data items subject to its memory capacity. The objective of cache admission control module is to store more distinct data items in the given cache space. This reduces the data traffic across the network since more number of requests can be served from the neighboring nodes. The cache consistency module is to maintain cache consistency, ie, ensuring that each node caching the data item is aware of the data update at the source. Due to the limited cache space, replacement policies decide which items should be evicted from the existing cache to make way for new ones. In this paper we focus on cache placement and resolution.

Towards the goal of improving the performance in cooperative cache, we propose a novel cache resolution scheme which uses a split table approach to resolve the cached data request which in turn reduces latency and message overhead. More precisely, our algorithm splits the neighboring nodes in to two sets based on communication range. The mobile devices keep the information about the neighboring nodes in two tables, top level nearest node cache table and next level nearest node cache table. The searching time and the request messages flooded in to the network is reduced by this method. Moreover, due to simplicity, this approach does not have any additional overhead in implementation. From the evaluation results we state that our proposal yields significant improvements compared to the previous solutions.

The remainder of the paper is structured as follows. Section 2 investigates the problem of cache resolution, Section 3 reviews the related works in cache resolution, Section 4 presents the system model and assumptions made and detail the proposed cache resolution, Section 5 describes simulation environment and metrics, section 6 gives the detailed analysis of results and section 7 concludes the paper.

2. Cache Resolution

Cache resolution in ad hoc networks constitutes an issue that several researchers are focusing on. There are a few interesting studies on this subject. Two basic mechanisms used to discover cache in ad hoc networks are broadcasting and cluster based approach. The first method is based on request/response message exchanges in which a user interested in discovering a cache broadcasts a discovery request containing information on the service requested and the node offering the data sends a reply. In the second method, a node known as cluster head is used to temporarily register the information about the cached data in the neighboring nodes. A node searching for data will first look in to the cluster head node to locate the data and then directly send a request to the corresponding node.

Although first method is the simpler version, it relays on flooding to broadcast the data request. Flooding increases network contention and overhead when the network density is high. In contrast, cluster based approach has been best owed with the features of reduced overhead by having a coordinator node which manages the cache resolution. The neighboring node which possesses the data can be easily found out checking the lookup table maintained by the cluster head. The disadvantage of this approach is that group maintenance is difficult due to the mobility of nodes. The control node may get disconnected which causes excessive overhead [4].The number of entries in the look up table increases when the network density is high. To maintain the correct status of the network, these tables must be frequently updated. This involves information exchange between the nodes which in turn increases the traffic overload in a dense network.

To circumvent these drawbacks we designed a cache resolution protocol among the mobile nodes. Our approach is based on a distributed cache discovery algorithm and reduced cache table entries to find the location of the requested data. An advantage of using this technique is that it reduces network traffic as flooding is not used. This technique also avoids the drawback of group maintenance by having a distributed approach. To reduce the number of entries in the look up table we split the tables in two based on transmission range.

3. Related Work

The main challenge in cooperative cache is to instantly find a cache containing the desired data. To facilitate data discovery in cooperative cache, some schemes [2][3][5][6][7] based on broadcasting and cluster based approach [9][12] is proposed in literature. These approaches differ in how the data request is resolved to find the location of the data stored in the neighboring nodes. In broadcast based approach the mobile nodes broadcast the request to find a node with the required data. On the other hand in cluster based approach, some mobile nodes are selected as information coordinators which deal with the task of finding the location of the required information. We can also find some discovery mechanism [8] that doesn't comes under this category.Below, we describe some representative cache resolution strategies for cooperative caching from the different groups mentioned above.

Aggregate caching scheme proposed in [5] tries to increase the data accessibility in an Internet based mobile network. A broadcast based information search algorithm called simple search is used to locate the required data item. Whenever a mobile nodes needs some data the request is broadcasted to its adjacent nodes .Upon receiving the broadcast request, the adjacent nodes replies to the request if it has already cached the data, otherwise the request is forwarded to its neighbors until it is acknowledged by an access point or some other nodes which have the requested data. Flooding is the technique used for broadcasting. This algorithm sets a hop limit for the request packet to reduce the traffic in the network. A caching technique which uses cluster based approach can be seen in [9], in which a coordinator node maintains the cluster cache state information of different nodes within its cluster domain. If there is a local cache miss, the coordinator node will find whether the data item is cached in other clients within its home cluster. Another approach for data discover other than the mentioned schemes can be seen in [4] and [10].In [4] a distributed group based approach in which each node maintains a group table and self table. Whenever a data request comes in it first checks in its self table, if data is not found checking is done in the group table which stores the data id and the node which posses the data. Here the mobile nodes interchange information about where the data is present in the network.

In [10], a cache resolution technique based on adaptive flooding broadcast is used for searching data in the network. According to this scheme a mobile node uses three schemes; adaptive flooding, profile-based resolution and road side resolution. In adaptive flooding, a node uses constrained flooding to search for items within the neighborhood. In profile-based resolution, a node uses the past history of received requests. In road side resolution, forwarding nodes caching the requested item, reply to the requests instead of forwarding them to the remote data source.

4. SYSTEM MODEL

4.1. Basic Assumptions

The system environment is assumed to be a mobile ad hoc network which includes a collection of nodes n_i (i=1,2,......N) that communicate with each other in a coordinated manner and $V_i \subseteq \{n1,n2,.......nN\}$ such that V_i is the set of neighboring nodes for a node 'n'. Data is originated in the server which is assumed to be in a fixed position. Mobile hosts are identified as <Hostid, Name> for $1 \leq i \leq N$, where N is the density of the network which is decided by the user. Each data item is uniquely identified by means of data item id D_i for $1 \leq i \leq n$ where 'n' is the size of the data base. Each node n_i maintains a local cache of data items which it frequently accesses. Whenever a node receives a request for the data item, local cache is checked. If the item is found the request is satisfied from the local cache, otherwise the request is forwarded to the neighboring nodes. The cache uses TTL based consistency.

4.2. Proposed Scheme

The objective of our cache resolution scheme is to reduce the overall access cost by reducing searching time and data traffic. Most of the current cooperative caching technique uses the entire list of one hop neighbors to find the cached data location. This will place unnecessary load on each node because they need to search all the entries in the table or broad cast a request to all of its neighbors to find the required data. The complexity of data discovery increases with the number of nodes in the network. In this paper we consider a cache resolution technique which uses a split table approach. In the given network each node N_i maintains a table nct_{i1}, whose entries represent a set of top level neighbors nei_1 ,which are located inside a wireless coverage of WR_1.The second table nct_{i2} represents a set of next level neighbors nei_2 ,that are located in a wireless range of WR_2 .

The table entries in each node of an ad hoc network are maintained as follows. The nodes coming under the wireless communication range WR_1 and WR_2 are taken and the distance between the current position and the node position is calculated using the Euclidian distance [5].When a node is farther away from the given node N_i the Euclidian distance will be more. The nodes are arranged on the basis of distance, signifying the first entry to be closest one .The correctness of the entries in the table is ensured by periodically updating the neighbor node information present in the tables. The distance of each node from node N_i is determined by using the Euclidian distance. For cache discovery, the top level neighbors are initially checked, and if we are not able to retrieve the data, request packet is forwarded to the nodes in the range WR_2. By this the redundancies in flooding is reduced. The communication cost is also reduced by avoiding multiple transmissions.

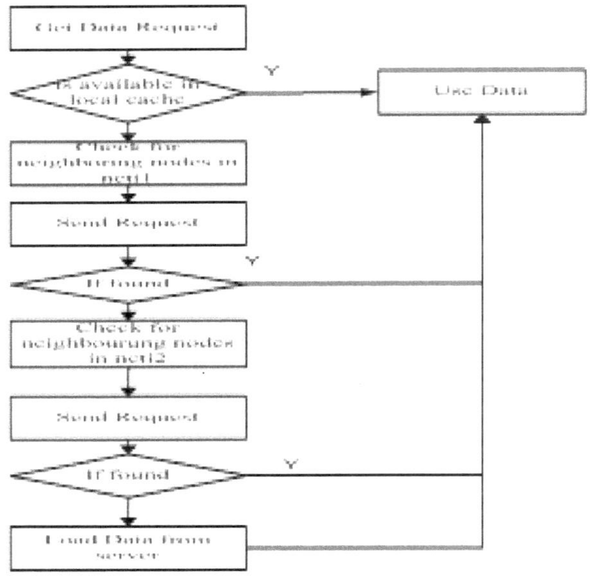

Fig.1 The cache resolution process

4.3. Cache Admission and Replacement

In cooperative caching nodes share the cache contents of neighboring nodes to utilize the full advantage of caching. The available cache replacement mechanisms for ad hoc network can be categorized in to coordinated and uncoordinated depending on how replacement decision is made [15]. In uncoordinated scheme the replacement decision is made by individual nodes. Coordinated cache replacement considers the information present in neighboring nodes for replacement.

In order to improve the content diversity in the cooperative cache, our scheme does not cache any data coming from the neighboring nodes. This increases the availability of information for the user, as more data items are cached and also avoids additional request to the server. Previous studies [16] have shown that the requests for smaller objects are more compared to bigger objects, so the probability of achieving high hit rate is increased if we store more number of small objects. In our proposed model we set a threshold value for the size of the data item admitted to cache. The threshold value is set as 50% of the total cache size. Any object bigger than this threshold isn't brought in to cache. The cache replacement is based on a key based algorithm, which takes in to account the inter arrival time of recent references, size and consistency for page replacement.

5. Simulation Set-up and Methodology

We have developed a simulation model in JAVA. In our simulation, nodes are randomly placed in an area of 800X800 m^2. Each node is identified by a node id and a host name. The data server is implemented as a fixed node in the simulation area. The data server contains all the data items requested by the mobile nodes. The size of each data item is uniformly distributed between s_{min} and s_{max}. The nodes in the network move randomly based on a random path. Simulation starts with each node having empty caches and, then, iteratively caches the data item in each node's cache and subsequent request for data are served from the cache. The nodes that generate data request are selected randomly and uniformly. The time interval between two consecutive queries generated from each node/client follows an exponential distribution with mean of 10sec. Each mobile node generates a single stream of read only queries. After a query is sent out, the client does not generate new query until the pending query is served. The data access pattern follows a Zipf distribution [14] with a skewness parameter □ as 0.8.

5.1. Simulation Parameters

Table.1 Simulation Parameters

Parameter	Value
Simulation Time	3600sec
Simulation Area	800X800 m^2
Database	1000 items
Cache size	20 -70 % of total database size
S_{min}	1
S_{max}	10
Nodes Density	20-70
Mobility model	Random waypoint
Transmission Range	500m
Speed of the mobile host	1-10
Pause time	200sec
Mean query generation time	10s

5.2. Evaluation Metrics

Our evaluation metrics include message overhead and cache hit ratio. The evaluation of these parameters is done by varying the number of nodes in each simulation run. The cache size is taken as 50% of the total size of the database. The hit ratio is defined as the percentage of requests that can be served from previously cached data. Message overhead is the overhead messages needed to manage the cache discovery process in cooperative cache.

6. Experiments and Results

We compared the performance of our method with broadcasting for different node densities. Fig 2-3 shows the different performance comparison of two schemes, cooperative caching with split table approach (ccs) and cooperative caching with broadcasting (ccb), as a function of different node densities. Fig 2 shows the message overhead for both schemes under different node densities. The figure show that ccs outperforms ccb at all node densities. As the node density increases, the difference become more significant which implies that ccs can benefit from larger node density. Fig 3 depicts the comparison of cache hit ratio for different node densities. From the figure we can see that the cache hit ratio for ccs and ccb perform quite closely. The relative performance of cache hit ratio remains relatively stable for higher network densities.

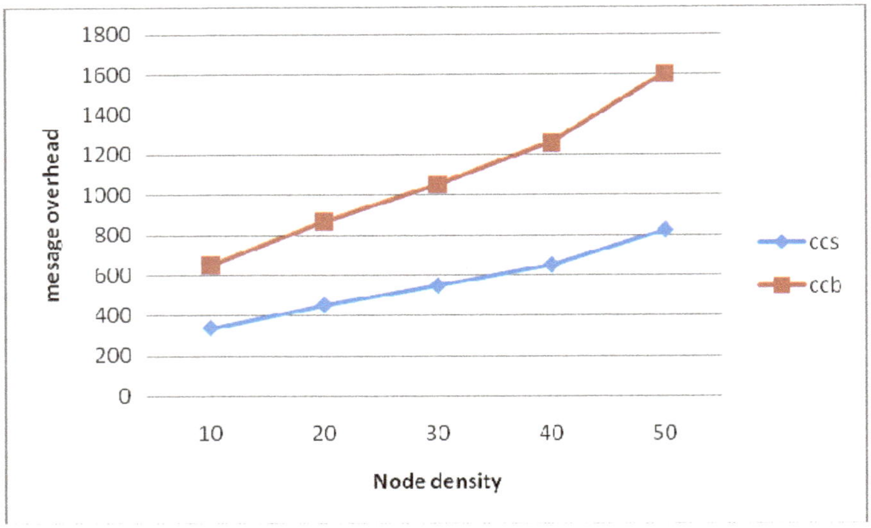

Fig. 2. Message overhead for different node densities.

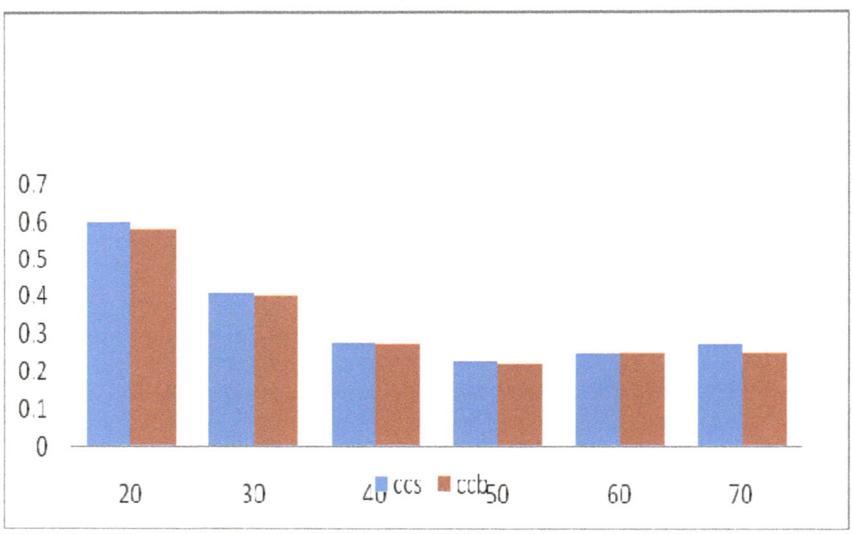

Fig .3 Cache Hit Ratio at different node densities

7. Conclusion

In this paper we addressed the problem of cache resolution technique for peer to peer cooperative caching in ad hoc networks. The objective of our problem was to minimize the number of messages flooded in to the network, which in turn reduces the communication cost and bandwidth utilization. We designed a data discovery process based on split table approach. We evaluated the performance our algorithm with broadcasting approach through extensive simulations. Experimental results show that the proposed cache resolution algorithm can significantly reduce the message overhead when compared to broadcasting.

REFERENCES

[1] N Chand, RC Joshi, M Misra, Cooperative caching in mobile ad hoc networks based on data utility. Mobile Inf Syst. 3 (1), 19 (2007).

[2] J Zhao, P Zhang, G Cao, CR Das, Cooperative caching in wireless P2P networks: design, implementation and evaluation. IEEE Trans Parallel Distributed Syst. 21(2), 229 (2010).

[3] Joonho Cho, Seungtaek Oh, Jaemyoung Kim, Hyeong Ho Lee, Joonwon Lee, Neighbor caching in multi-hop wireless ad hoc networks, IEEE Communications Letters, Volume 7, Issue 11, 525 – 527(2003).

[4] Yi-Wei Ting,Yeim-Kuan Chang. A novel cooperative caching scheme for wireless ad hoc networks: Groupcaching. In NAS '07: Pro-ceedings of the International Conference on Networking, Architecture,and Storage,(2007).

[5] S Lim, WC Lee, G Cao, CR Das, A novel caching scheme for improving Internet-based mobile ad hoc networks performance. Ad Hoc Netw. 4 (2),225 (2006).

[6] L Yin, G Cao, Supporting cooperative caching in ad hoc networks. IEEE Trans Mobile Comput. 5 (1), 77 (2006)

[7] M Fiore, F Mininni, C Casetti, DF Chiasserini, To cache or not to cache? In Proceedings of the IEEE Conference on Computer and Communications (INFOCOM 2009) , Rio de Janeiro, Brazil 235 (2009).

[8] B Tang, H Gupta, SR Das, Benefit-based data caching in ad hoc networks.IEEE Trans Mobile Comput. 7 (3), 289 (2008)

[9] Denko, M.K., Tian, J.,Cross-Layer Design for Cooperative Caching in Mobile Ad Hoc Networks, Proc .of IEEE Consumer Communications and Networking Conf(2008).

[10] Y. Du,Gupta, S. K. S.:COOP-A cooperative caching service in MANETs, in Proc. Joint Int. Conf Autonomic and Autonomous Systems and Int. Conf. Networking and Services,. 58-63(2005).

[11] Dan Hirsch, Sanjay Madria, A Resource- Efficient Adaptive Caching Scheme for Mobile Ad-Hoc Networks.29th IEEE International Symposium on Reliable Distributed Systems (2010).

[12] González-Cañete et al : A cross layer intercep tion and redirection cooperativ e caching scheme for MANETs .EURASIP Journal on Wireless Communications and Networking 2012.

[13] Tiance Wang , Pan Hui, Sanjeev R. Kulkarni, Paul Cuff, Cooperative Caching based on File popularity Ranking in Delay Tolerant Networks, Proc. Of ExtremeCom , 2012, Zürich, Switzerland

[14] L. Breslau, P. Cao, L. Fan, G. Phillips, and S. Sheker, Web Caching and Zipf-Like Distributions: Evidence and Implications, Proceedings of IEEE INFOCOM, pp.126-134,March 1999.

[15] Preetha Theresa Joy and K. Polouse Jacob, Cache Replacement Strategies for Mobile Data Caching ,International Journal of Ad hoc, Sensor & Ubiquitous Computing, August 2012, Volume 3, Number 4.

[16] S. Jin, A. Bestavros, GreedyDual Web caching algorithm: exploiting the two sources of temporal locality in web request streams. Computer Communications, vol. 24, no. 2, Feb. 2001, pp. 174-83.

MULTIPLE OPTIMAL PATH IDENTIFICATION USING ANT COLONY OPTIMISATION IN WIRELESS SENSOR NETWORK

Aniket. A. Gurav[1], Manisha. J. Nene[2]

Department of Applied Mathematics[1]
Department of Applied Mathematics and Computer Engineering [2]
Defence Institute of Advance Technology[12], Pune, India 411 025

ABSTRACT

Wireless Sensor Network WSN is tightly constrained for resources like energy, computational power and memory. Many applications of WSN require to communicate sensitive information at sensor nodes SN to Base station BS. The basic performance of WSN depends upon the path length and numbers of nodes in the path by which data is forwarded to BS. In this paper we present bio-inspired Ant Colony Optimisation ACO algorithm for Optimal Path Identification OPI for packet transmission to communicate between SN to BS. Our modified algorithm OPI using ACO is base-station driven which considers the path length and the number of hops in path for data packet transmission. This modified algorithm finds optimal path OP as well as several suboptimal paths between SN & BS which are useful for effective communication.

KEYWORDS

WSN, Ad-Hoc Network, ACO, Optimal path Identification, Multiple optimal paths

1. INTRODUCTION

A WSN is an ad-Hoc network composed of hundreds or even thousands of SN. These SN are capable of sensing. SN's are battery powered. Every time replacing or recharging of battery is not possible. The SN's have to perform tasks such as object monitoring & tracking, detecting the presence of certain objects, event monitoring, data fusion & localization. These tasks of SN generate a vast amount of data. The SN has to forward this data to BS (sink node).It is found that SN consumes lot of energy while communicating this information so over all network lifetime will increase if WSN consumes minimum energy in communication. The reporting between SN & BS consumes energy based on the type of communication protocol, communication path & number of hops between SN and BS. This requires finding out optimal path OP between BS to SN which will be useful in efficiently forwarding data, reducing delay, evenly distributing traffic load & power consumption and communication overhead in WSN.

In our work we are focused on identifying optimal and suboptimal communication path between SN and BS.

There are various routing protocols like Direct diffusion [9], LEACH (Low Energy Adaptive Clustering Hierarchy) [8], SPIN (Sensor protocols for Information via Negotiation) [8] [10].The Shortest path finding is the backbone for the routing algorithms in WSN. So it is a graph based problem to find out the shortest path between two vertices. As WSN is resource constrained it is not desirable to use the conventional shortest path finding techniques.

So we should look at soft computing techniques. It can help us in dynamically changing environment for setting communication paths in WSN. ACO is a soft computing technique for Optimisation. This method is having characteristic like achieving global optimization through local interaction, high degree of self organization and adaptiveness.

1.1 Motivation

Routing in WSN is challenging due to limited capabilities of SN, dynamically changing communication network environment and network topology, abrupt failure of SN etc. To handle these many routing techniques exist, such as data centric, location based and hierarchical routing. Finding shortest path in WSN during routing can help to optimise communication and computation overhead.

In target or object tracking, surveillance vast amount of data is generated. This data should be forwarded to BS with an optimal path to process it in the real time. A Greedy algorithm like Dijkstra's algorithm, Bellman Ford and Dynamic programming algorithm are useful in shortest path finding but it is having high computational complexity. Greedy algorithm doesn't give a guarantee of finding the globally OP and these techniques is helping to identify a single static shortest path but the only single shortest path is not desirable for WSN.

Bio-inspired shortest path algorithm using ACO proves to be efficient in the OP detection. By studying traditional ACO we found its following drawback for WSN which are possible to overcome. Firstly ACO based routing [1] [4] algorithm which is the present finds path by creating ants in the form of data packets, Data packets are transmitted between each node i.e. peer to peer. This increases the communication overhead which is likely to result in increased network traffic. Also packet collision and packet loss can result decrease in efficiency.

Existing routing protocols can be improved by considering the path length and the number of nodes in path as a critical parameter to detect the optimal path between SN to BS. Secondly the traditional ACO [11] considers only path length not the number of hops in the path and it is not guaranteed to lead the optimal solution every time. Considering these challenges we have modified the Traditional ACO in an appropriate way .Our work overcomes above mentioned issues & BS computes the OP and communicates it with the SN. So in our approach path discovery is done prior to the data transmission.

1.2 Overview of Our Work

In our work we are proposing BS driven OPI. Here BS is made aware (directly or indirectly) about the all SN their position and topology of the network. So problem of finding the optimal path now becomes graph based problem.

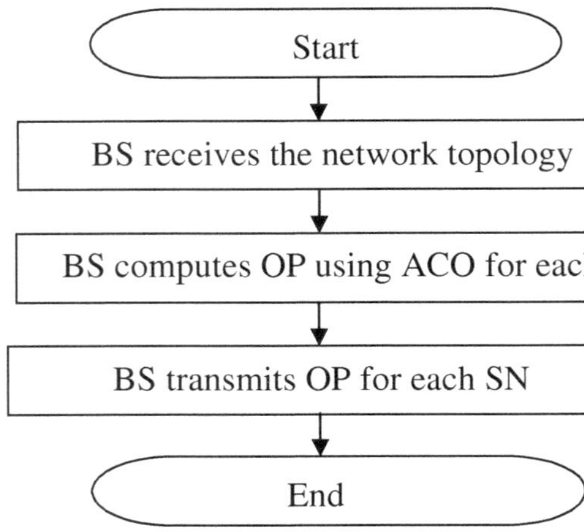

FIG. 1 SHOWS OVERVIEW OF OUR WORK

Here BS calculates OP using ACO & communicates that path with respective SN. So in 1st step BS made aware about the network topology. In 2nd OP is identified. Then in the 3rd step OP is communicated with the respective SN. After all these steps are done then any SN who wants to send data to BS can use above calculated OP.

1.3 ASSUMPTIONS & EXPERIMENT PARAMETERS

For applying ACO for finding OP our assumptions are as follows.

a) Assumption at BS & SN

BS is aware about co-ordinate of all SN present and so it is also aware about the topology of the entire WSN. BS is responsible for the entire computation of ACO. These are assumed about BS. In case of SN it is assumed that SN's are already localized. All SN's are static. SN's cannot be recharged. Communication range of all SN is same.

b) Assumption for OPI Using ACO

Traditional ACO is explained in this paper, for our experiment we have taken the value of parameters $\alpha=1$, $\beta=3$, $\rho=0.2$. In our experiment WSN is treated as a graph. All the SN's which are in communication range of each other are represented as adjacent nodes in the graph. θ is no of nodes (hops) in the path from source to destination. Ω a is a parameter which manages relative importance of path length & ψ controls the importance of nodes in path (equation 6). The value of these two variables is 3.

2 ROUTING USING ACO

Routing using ACO is a complex task due to resource constrained WSN. In routing SN node route data to BS using the most efficient path. In ACO based approach behaviour of real ant searching for food through pheromone deposition helps to find the optimal path.

2.1 Routing using Traditional ACO

When ants trace out a path from their nest to a food source, ant drop chemical called pheromone on the path. On the shorter path more pheromone gets accumulated. This is because shorter paths accumulate pheromone deposits at a faster rate. Suppose each ant starts from the source "s" to destination "*d*", it tries to find the shortest path between these nodes. At each node "*i*" ant "*k*" decides to visit the next node "*j*" based upon the probability given by formulae in equation 1.

$$p_{ij}^{k} = \begin{cases} \dfrac{[\tau_{ij}]^{\alpha} * [\eta_{ij}]^{\beta}}{\sum [\tau_{ij}]^{\alpha} * [\eta_{ij}]^{\beta}} & \forall j \in N_j \ \& \ j \notin M^k \\ 0 & \textit{otherwise} \end{cases} \quad (1)$$

Where τ_{ij} is the pheromone concentration on edge between node "*i*" to node "*j*". η_{ij} is the value of heuristic related to path length, α and β are two parameters which controls the relative importance of pheromone trail and heuristic value related to path length. N_j is set of SN. M^k Is the memory of ant "k" The heuristic value related to path length is

$$\eta_{ij} = \frac{1}{l(i, j)} \quad (2)$$

Where $l(i, j)$ is the edge length between nodes "i" and "*j*".

After each iteration "*t*", ants "k" deposit quantity of pheromone which is given by

$$\Delta \tau^{k}(t) = (1 / j^{k}(t)) \quad (3)$$

Where $j^{k}(t)$ is the length of the path from source to destination traversed by ant "k". Total Amount of pheromone quantity on edge "*i*" to "*j*" is given by the equation

$$\tau_{ij}(t) \leftarrow \tau_{ij}(t) + \Delta\tau_{ij}(t) \quad (4)$$

With increase in number of iteration/time parameter the pheromone starts evaporating. A control coefficient $\rho \in [0,1]$ decides the amount of pheromone on each edge at any specific iteration which is given by the equation

$$\tau_{ij}(t) \leftarrow (1-\rho)\tau_{ij}(t) \quad (5)$$

As the number of iteration increases pheromone on shortest path becomes more as compared to other paths present in the WSN. So the number of ants gets attracted to shorter path. Finally pheromone concentration on this path becomes higher and algorithm converges.

2.2 OPI using ACO

Steps1 to 7 describes the working of our proposed algorithm. Our proposed algorithm is modified according to the needs of energy constrained WSNs. The proposed algorithm considers above mentioned two parameters path length and number of hops in path to find out OP.

Steps

Step 1: When the BS starts computing shortest path, it decides one destination node in

WSN. An ant is created which will generate a path from BS to that destination node.

Step 2: Ant "*k*" at node "*i*" selects the next node "*j*" using formula in equation 1.
Here "*j*" is one of the adjacent node of "*i*". An ant "k" has more probability to choose the node with larger values of p_{ij}^{k} the next node selected is stored in memory of ant k denoted by (M^{k}).

Step 3: If any ant visits the node which is already visited by the same ant then that ant Is discarded from path calculation.

Step 4: Step 2 and 3 are repeated until ant "k" finds its destination node or discarded.

Step 5: Step 1 to 4 is repeated for all ants.

Step 6: When all ants complete above procedure pheromone on the edge is updated which is given by equation 6.

$$\tau_{ij}(t) \leftarrow \tau_{ij}(t) + ((\frac{1}{\theta^{\Psi}}) \times (1 + (\frac{1}{j^{k}(t)^{\Omega}})) \quad (6)$$

Step 7: Then the evaporation of pheromone is calculated by equation 5.

Step 8: Now the path traversed by all ants is compared. The best optimal path is selected Among them by comparing the number of hops & distances. This optimal The path is then deleted from the graph. Again same procedure is repeated, this time result will be a different optimal path than the previous one because that path is already deleted.

As the above steps are iterated, BS comes to know about the optimal & suboptimal paths between it and the destination. Similarly the BS can calculate distance between it and all other SN.

2.3 RESULT

This section compares the result of optimal path found using modified ACO and traditional ACO. Figures 2,3 and 4 shows deployed WSN and its communication network. The nodes which can directly communicate is shown by edges. The distance between corresponding SN is indicated near the edges.

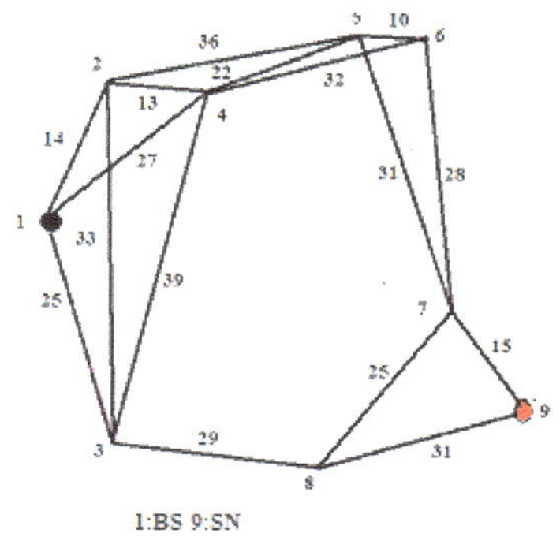

Fig. 2 Wireless Sensor Nodes and its network

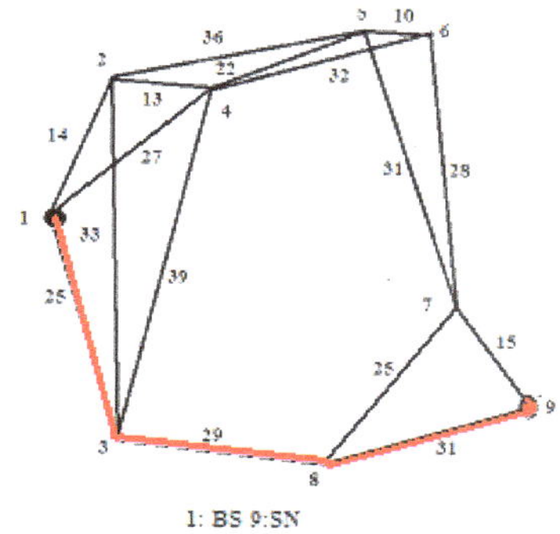

Fig. 3 OP using modified ACO

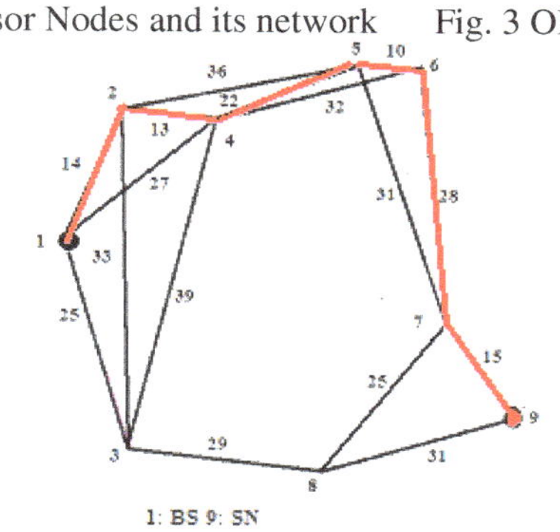

Fig. 4 Optimal Path using traditional ACO

In the above scenario of figure 1 node 9 senses some event and it want to forward that information to node 1. Node 1 which is BS is aware about co-ordinates and the topology of the entire network .It implements proposed algorithm and finds the optimal paths between SN 1 and 9. This step is repeated for all nodes and optimal path found is communicated with all SN. So all SN including 9 is now aware about the optimal path between itself to BS. In this way finding the optimal paths is done. So every SN know how to forward data to BS. The initial pheromone concentration on all edges is as shown in table number 1.The left most columns and top row denote the node numbers. Here N in top most left block in table 1 stands for node each edge is initialised with pheromone quantity 0.00001. This is initialisation necessary to begin the execution of ACO. This pheromone quantity is kept as small as possible because it should not affect the final result. As the ants traverse the path, pheromone concentration on that edge start increasing. Finally more pheromone gets deposited on the path with the shorter length, which is overused. This converges to the optimal path. After implementing the modified algorithm the pheromone concentration in a 5th iteration is shown in table 2. Here left column and top row denote the node number and N in top most left block in table 2 represents the sensor node.

Table 1 Pheromone value

N	1	2	3	4	5	6	7	8	9
1	0	0.00001	0.00001	0.00001	0	0	0	0	0
2	0.00001	0	0.00001	0.00001	0.00001	0	0	0	0
3	0.00001	0.00001	0	0.00001	0	0	0	0.00001	0
4	0.00001	0.00001	0.00001	0	0.00001	0.00001	0	0	0
5	0	0.00001	0	0.00001	0	0.00001	0.00001	0	0
6	0	0	0	0.00001	0.00001	0	0.00001	0	0
7	0	0	0	0	0.00001	0.00001	0	0.00001	0.00001
8	0	0	0.00001	0	0	0	0.00001	0	0.00001
9	0	0	0	0	0	0	0.00001	0.00001	0

2.3.1 Optimal Path Identification from Pheromone Table

OP is detected from pheromone concentration which is given in table 2. For choosing the optimal path we just have to look at the edges (table block) with a maximum pheromone concentration starting from sensor 1 (in row 1) till we reach sensor 9. So from 1st node pheromone concentration is maximum in the 3rd column (3rd column in 1st row) which is 0.0001059. So edge from 1 to 3 is added. Then from the 3rd sensor (from 3rd row) look for maximum pheromone value. 8th column value is 0.0001138 which is the maximum. So add edge from 3 to 8. With the node 8, the maximum pheromone concentration is 0.001153, so edge from 8 to 9 is added. By proceeding this way one can find a path. In Figure 3 the path with the maximum pheromone concentration is marked with red colour. This path is optimal path. From table 2 it is clear that the pheromone concentration on the path 1 to 3, 3 to 8, 8 to 9 is higher. So our optimal path is the 1-3-8-9.The length of this path is 85. We can also get a parallel path if we initially choose 2nd highest node from node 1 & then keep choosing highest pheromone node. The other parallel path proposed by the implemented algorithm OPI using ACO is 1-2-4-5-6-7-9. This path is not optimal but it is sub optimal. This suboptimal path selection is useful when an optimal path for communication is required to be chosen between the communicating nodes. Especially, when one optimal path is over utilised and there is a need to conserve energy of nodes on that path. The length of suboptimal path is 102. Table 3 gives the number of ants choosing the sensor nodes. Where left column and top row denote nodes. Here N in top most left block in table 3 stand for node. The number of Ants choosing a path is proportional to the pheromone concentration on the path. Hence many times the path found in Ant table is the same as the result given by the pheromone value. Instead of choosing edge with the highest ant count, if 2nd highest edge is taken at node 1, then we get another suboptimal path. That path is 1-2-4-5-7-9.This alternate path is very useful as using the same OP will reduce the energy level of the SN in the path. This enables a dynamic detection of alternative optimal path for communication between the communicating sensor nodes.

Table 2 Pheromone value

N	1	2	3	4	5	6	7	8	9
1	0	0.0000353	0.0001059	0.000015	0	0	0	0	0
2	0.0000353	0	0.0000185	0.0000317	0.0000166	0	0	0	0
3	0.0001059	0.0000185	0	0.0000205	0	0	0	0.0001138	0
4	0.000015	0.0000317	0.0000205	0	0.0000221	0.0000158	0	0	0
5	0	0.0000166	0	0.0000221	0	0.0000227	0.0000216	0	0
6	0	0	0	0.0000158	0.0000227	0	0.0000208	0	0
7	0	0	0	0	0.0000216	0.0000208	0	0.0000145	0.000309
8	0	0	0.0001138	0	0	0	0.0000145	0	0.001153
9	0	0	0	0	0	0	0.000309	0.001153	0

Table 3 Ant count

N	1	2	3	4	5	6	7	8	9
1	0	42	63	15	0	0	0	0	0
2	0	0	11	35	20	0	0	0	0
3	0	12	0	12	0	0	0	69	0
4	0	14	16	0	21	14	0	0	0
5	0	1	0	3	0	22	28	0	0
6	0	0	0	2	12	0	23	0	0
7	0	0	0	0	3	7	0	13	36
8	0	0	5	0	0	0	8	0	69
9	0	0	0	0	0	0	0	0	0

2.4 Comparison of Results with Traditional ACO

Traditional ACO is already described in this paper. The comparison of results of these two algorithms is given below. The initial pheromone concentration of Traditional ACO is in the table 1. After 5th iteration the pheromone concentration is as shown in table 4 .From table 4, it is clear that pheromone concentration is more on the path 1-2-4-5-6-7-9.The length of this path is 102. So ants will start preferring this path. In the result section 6 by modified ACO we have found an optimal path which is equal to 85.So it proves that our modified algorithm outperforms the traditional ACO. Figure 4 shows path found by traditional ACO with red colour.

Now the same method is applied to more complex scenario shown in figure 5. In this scenario number of SN are more & they are densely distributed. Here aim is to find the Multiple Optimal Paths MOP between BS node 1 and SN 30. The modified ACO method which is described previously is used for finding the single Optimal Path. While the optimising number of hops and path length from SN to BS is considered. So these OP are having less path length as well as number of hops. After each iteration, if ants are converged to the single path then that path is selected as the optimal path. As our aim is to find the multiple optimal paths the path which is discovered in previous iteration is deleted from the graph so ants in this iteration will not be able to travel through that path. Deletion of the optimal path in the previous iteration causes ACO to search for the new optimal path in this iteration. In this way several optimal paths are discovered.

Table 4 Pheromone value for Traditional ACO

N	1	2	3	4	5	6	7	8	9
1	0	11.462	0.3922	0.4440	0	0	0	0	0
2	11.462	0	0.6559	10.930	0.3302	0	0	0	0
3	0.3922	0.6559	0	1.066	0	0	0	1.564	0
4	0.4440	10.930	1.0663	0	4.457	1.422	0	0	0
5	0	0.3302	0	4.457	0	13.41	1.257	0	0
6	0	0	0	1.422	13.414	0	3.530	0	0
7	0	0	0	0	1.2572	3.53	0	0.539	8.275
8	0	0	1.564	0	0	0	0.539	0	1.799
9	0	0	0	0	0	0	8.275	1.799	0

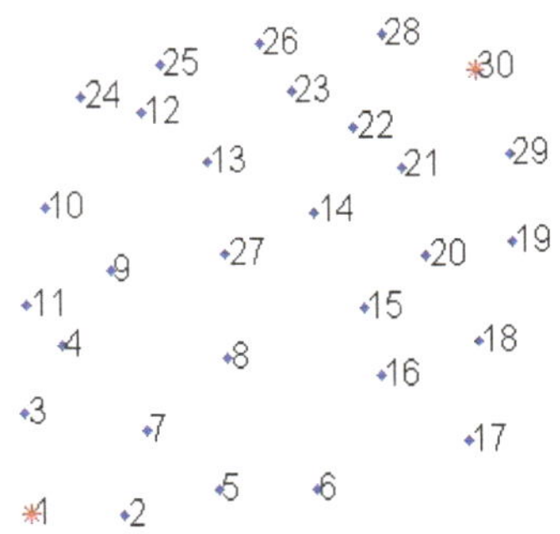

Fig. 5 Wireless sensor nodes and its network

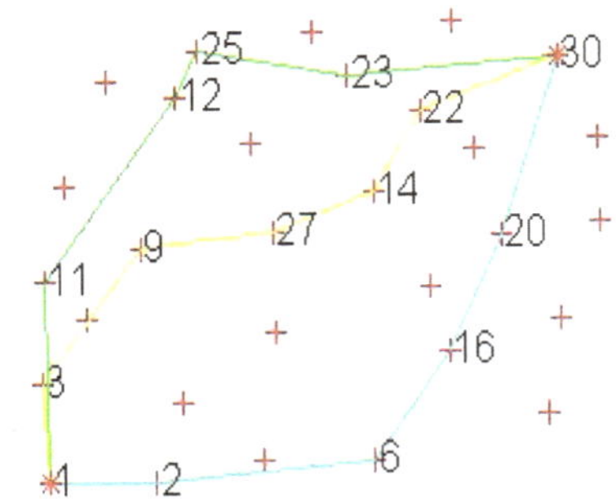

Fig. 6 MOP using modified ACO

Figure 6 shows collection of different optimal and suboptimal paths with different colours.This algorithm is applied for number of SN between 50 to 100. Every time we have got Multiple Optimal paths as described.

3.CONCLUSION

Finding the optimal path in dynamically changing resource constrained WSN is challenging. Our work proposes an approach to identify multiple optimal paths for communication between SN to BS. Existing greedy approach is static in providing optimal path. In our computational approach and communication overhead is reduced as BS takes the responsibility of computation & optimal path is calculated based on pheromone concentration. Multiple optimal paths between SN and BS are useful in forwarding data, efficiently distributing the load. This increases the throughput and increases the network lifetime.

REFERENCES

[1] Selcuk Okdem and Dervis Karaboga, "Routing in Wireless Sensor Networks Using an Ant Colony Optimization (ACO) Router Chip", Sensors 2009, 9, 909-921;doi: 10.3390/s90200909

[2] P.K.Singh, Narendra pal singh, "Data Forwarding in Ad-hoc Wireless Sensor Network Using Shortest Path Algorithm", Volume 2, No. 5, May 2011 Journal of Global Research in Computer Science

[3] Adamu Murtala Zungeru, Li-Minn Ang, S.R.S. Prabaharan, and Kah Phooi Seng, "Ant Based Routing Protocol for Visual Sensors", A. Abd Manaf et al. (Eds.): ICIEIS 2011, Part II, CCIS 252, pp. 250–264, 2011. , Springer-Verlag Berlin Heidelberg 2011.

[4] Selcuk Okdem, Dervis Karaboga "Routing in Wireless Sensor Networks Using Ant Colony Optimization", Proceedings of the First NASA/ESA Conference on Adaptive Hardware and Systems (AHS'06) 0-7695-2614-4/06 2006 IEEE

[5] Dilpreet kaur, Dr.P.S Mundra, "Ant Colony Optimization: A Technique Used For Finding Shortest Path", International Journal of Engineering and Innovative Technology (IJEIT) Volume 1, Issue 5, May 2012

[6] Masaya Yoshikawa, Kazuo Otani, " Ant Colony Optimization Routing Algorithm with Tabu Search", Masaya Yoshikawa, Kazuo Otani.

[7] Jamal N. Al-karaki, Ahmed E.Kamal, "A Taxonomy of Routing Techniques in Wireless Sensor Network", in Sensor Network Protocols Taylor & Francis page 3-1 to 4-24 [8] Gizem, Aksahya & Ayese, Ozcan (2009) *Coomunications & Networks*, Network Books, ABC Publishers.

[8] W. R. Heinzelman, A. Chandrakasan, and H. Balakrishnan, "Energy-Efficient Communication Protocol for Wireless Microsensor Networks", proc. Hawaii International Conference on System Sciences, Vol. 8, pp. 1-10, Jan. 2000.

[9] C.Intanagonwiwat, R.Govindan, and D.Estrin, "Directed diffusion for wireless sensor network", IEEE /ACM transaction . Networking,11(1),2-16 ,2003

[10] J.Kulik, W.R.Heinzelman, H.Balakrishnan, "Negotion based protocol for disseminating information in wireless sensor network", Wireless Networks, 8, 169-185,2002

[11] M. Dorigo , G. Di Caro, "Ant colony optimization: a new meta-heuristic", Proceedings of the 1999 Congress on Evolutionary Computation, Proceedings, 1999

15

TWO HOP CONNECTIVITY FOR UNIFORMED RANDOMLY DISTRIBUTED POINTS IN THE UNIT SQUARE

Johannes Karlsson

Embedded Systems Lab, Umeå University, Sweden
johannes.karlsson@tfe.umu.se

ABSTRACT

Connectivity in ad-hoc networks is a fundamental, but to a large extend still unsolved problem. In this paper we consider the connectivity problem when a number of nodes are uniformly distributed within a unit square. We limit our problem to the one-hop and two-hop connectivity. For the one-hop connectivity we calculate the exact analytically solution. For the two-hop connectivity we calculate the lower and upper bound for connectivity.

KEYWORDS

Connectivity, Ad-Hoc networks

1. INTRODUCTION

In wireless ad-hoc networks it is important to know the probability to have a path between two nodes for a given node density. This can easily be found using simulations but no analytically solution is available. In this paper we consider the case when a number of nodes are uniformly distributed in a unit square. The reason why we use this distribution is to make it possible to compare the results with network simulators where the nodes often are uniformly distributed within a square. We will calculate the exact analytically solution for the one hop case, when there are only two nodes, and the upper and lower bound for the two-hop case.
Some related research has been done in this area. The theory of random geometric graphs is a rather realistic model for wireless networks. In this model n nodes are randomly distributed in a d-dimensional space with a common density f. Here the status of different edges in the graph is not independent. If Xi is close to Xj , and Xj is close to Xk, then Xi will be fairly close to Xk. Connectivity has been studied in this type of graphs, but only for very dense graphs [1]. Feng Xue et al. are considering the number of neighbors needed for connectivity of wireless networks [2]. The nodes are uniformly and independently distributed in a unit square. They have found the number of neighbors needed for the network to be asymptotically connected. This definition does not give the answer to the probability for connectivity as a function of the node density. Leonard E. Miller has addressed a similar problem as our for two-hop connectivity [3], but in this paper he assumes Gaussian distribution between the points. In our paper we use uniform distribution. The one-hop connectivity has been solved by Miller [4]. We include our solution for one hop connectivity in this paper since we use the same approach to solve two hop connectivity.

2. THE DISTRIBUTION FOR THE DISTANCE BETWEEN TWO RANDOM POINTS

Let two points, X and Y , be uniformly and independently distributed in the unit square, K (1). The ball, $B\tau$ (p), with center p and radius τ is defined in (2). Let Γ be the area of the part of the circle inside the unit square having the center in point X and a radius of τ. The probability for the distance between X and Y to be less or equal to τ is the same as the expected size of the area Γ, (3). The distance between the points can be between 0 and $\sqrt{2}$. It is however not interesting to consider connectivity for large values of τ. We are therefore only solving the probability density function for τ between 0 and 1.

$$X, Y \in U(K), |K| = 1 \tag{1}$$

$$B_\tau(p) = \{y : |y - p| \leq \tau\}, \; y, p \in \Re^2 \tag{2}$$

$$P(d(X, Y) \leq \tau) = P(Y \in B_\tau(X)) = E(\Gamma) \tag{3}$$

Depending on where in the unit square the point X (center of the circle) is located the size of the area of the circle is calculated in different ways. If the distance from point X to one edge is less than the radius then a segment of the circle will be located outside of the unit square. If the distance from point X is less than the radius from two edges then two circle segments will be outside of the unit square. Otherwise the entire circle will be within the unit square. We therefore divide the unit square into three different zones, see Fig. 1.

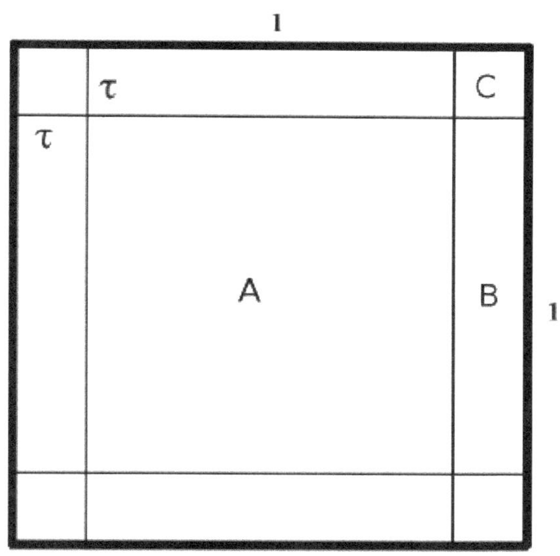

Figure 1. The different zones used in the unit square.

We then calculate the probability for point X to be located in each of the zones (4,5,6) and then the expected size of the area of the circle given that point X is located in each zone (7,13,19). The probability for point X to be located in each of the zones is given by:

$$P(X \in A) = (1 - 2\tau)^2 \tag{4}$$

$$P(X \in B) = 4\tau(1 - 2\tau) \tag{5}$$

$$P(X \in C) = 4\tau^2 \tag{6}$$

For zone A the entire circle is always within the unit square and the expected area of the circle is given by:

$$E(\Gamma | X \in A) = \pi\tau^2 \tag{7}$$

When point X is located in zone B a segment of the circle is outside of the unit square. We use the variable n for the distance between the edge and the center of the circle. This distance is uniformly distributed between 0 and τ.

$$n \in [0, \tau]$$

The size of the circle segment, M, as a function of n is given by substituting (8) and (9) in (10):

$$s = 2\tau \cos^{-1}(n/\tau) \tag{8}$$
$$c = 2\sqrt{\tau^2 - n^2} \tag{9}$$
$$M = (1/2)(\tau s - nc) \tag{10}$$

$$\begin{aligned} M &= g(n, \tau) \\ &= \tau^2 \cos^{-1}(n/\tau) - n\sqrt{\tau^2 - n^2} \end{aligned} \tag{11}$$

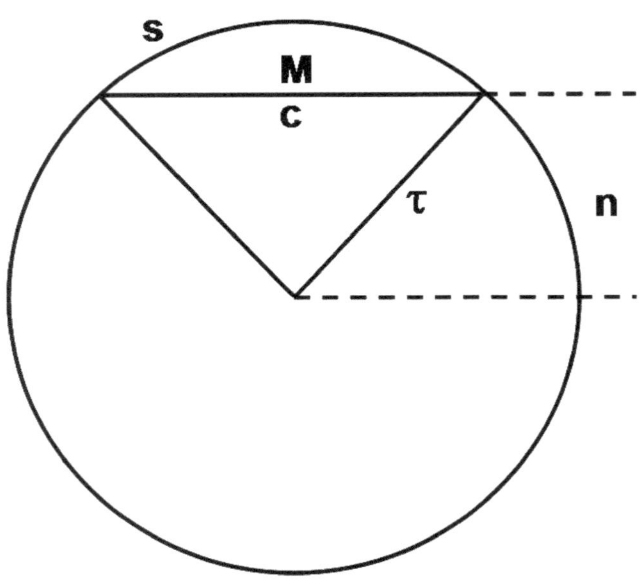

Figure 2. A segment of a circle.

The expected size of the circle segment outside of the unit square is given by:

$$\begin{aligned} E(M) &= E(g(n, \tau) f_n(n)) \\ &= \int_{n=0}^{\tau} \left(\tau^2 \cos^{-1}\left(\frac{n}{\tau}\right) - n\sqrt{\tau^2 - n^2} \right) \frac{1}{\tau}\, dn \\ &= (2/3)\tau^2 \end{aligned} \tag{12}$$

The expected size of the area inside the unit square when point X is located in zone B is given by:

$$\begin{aligned} E(\Gamma | X \in B) &= \pi\tau^2 - E(M) \\ &= (\pi - (2/3))\,\tau^2 \end{aligned} \tag{13}$$

For zone C the circle can be partly located outside of two edges. For each of the edges the average size outside of the edge is given by (12).
There will be an overlapping between these areas. This is the area L in Fig. 3. The position (Q, U) for the corner will be random and uniformly distributed (14). The area L will be a function of the position of the corner (15).

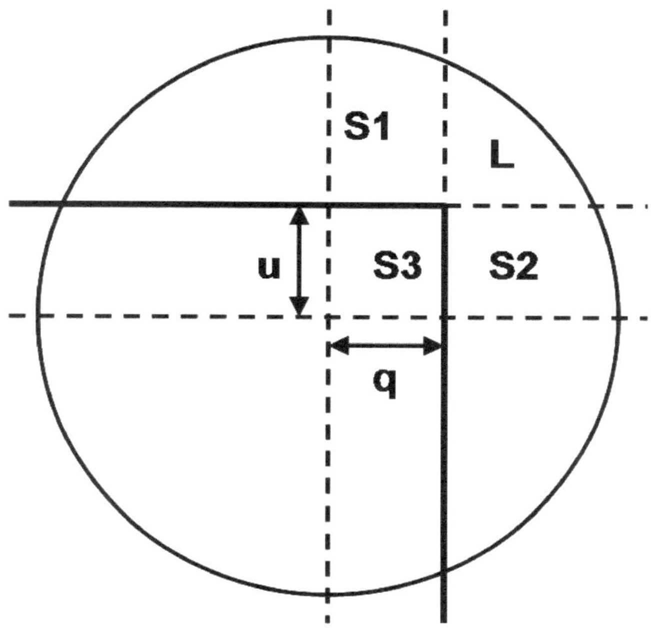

Figure 3. The overlapping area between two circle segments.

$$Q, U \in [0, \tau] \tag{14}$$

$$L = g(q, u)\, I_1(\sqrt{q^2 + u^2}) \tag{15}$$

$$I_1(i) = \begin{cases} 1, & i \leq \tau \\ 0, & i > \tau \end{cases}$$

We first calculate the function g(q, u) in (16). This is the area (S1 + L) + (S2 + L)+(S3)−(S1+S2+S3+L). The areas (S1+L) and (S2+L) are half circle segments and these areas are found by using (11). The expected size of L is then found in (18). The pdf for (q, u) is given in (17).

$$
\begin{aligned}
g(q, u) &= \frac{\tau^2 \cos^{-1}(q/\tau) - q\sqrt{\tau^2 - q^2}}{2} \\
&\quad + \frac{\tau^2 \cos^{-1}(u/\tau) - u\sqrt{\tau^2 - u^2}}{2} \\
&\quad + qu - \frac{\pi\tau^2}{4}
\end{aligned} \tag{16}
$$

$$
f_{qu}(q, u) = 1/\tau^2 \tag{17}
$$

$$
\begin{aligned}
E(L) &= \int_{q=0}^{\tau} \int_{u=0}^{\sqrt{\tau^2 - q^2}} g(q, u) f_{qu}(q, u)\, du\, dq \\
&= (1/8)\tau^2
\end{aligned} \tag{18}
$$

The expected size of the area inside the unit square when point X is located in zone C is given by:

$$
E(\Gamma | X \in C) = \pi\tau^2 - (E(M) - E(L)) \tag{19}
$$

$$
\begin{aligned}
&= \pi\tau^2 - \left(2(2/3)\tau^2 - (1/8)\tau^2\right) \\
&= \pi\tau^2 - (29/24)\tau^2
\end{aligned} \tag{20}
$$

The total contributed expected size from each of the three areas is given by multi- plying the probability for being in that area with the expected size when the center of the circle is located inside of each of the areas.

$$
\begin{aligned}
&E(\Gamma | X \in A) P(X \in A) \\
&= (1 - 2\tau)^2 \pi\tau^2 \\
&= \pi\tau^2 - 4\pi\tau^3 + 4\pi\tau^4
\end{aligned} \tag{21}
$$

$$
\begin{aligned}
&E(\Gamma | X \in B) P(X \in B) \\
&= 4\tau(1 - 2\tau)\left(\pi\tau^2 - (2/3)\tau^2\right) \\
&= 4\pi\tau^3 - 8\pi\tau^4 - (8/3)\tau^3 + (16/3)\tau^4
\end{aligned} \tag{22}
$$

$$
\begin{aligned}
&E(\Gamma | X \in C) P(X \in C) \\
&= 4\tau^2 \left(\pi\tau^2 - (29/24)\tau^2\right) \\
&= 4\pi\tau^4 - (29/6)\tau^4
\end{aligned} \tag{23}
$$

By adding (21), (22) and (23) we will have the expected area of a circle having radius τ with the center uniformly distributed inside a unit square. This is also the probability for two randomly uniformly distributed points to have a distance less or equal to τ. This is the cumulative probability distribution function (cdf) for the distance between to random points in the unit square. (24).

$$
F(\tau) = (1/2)\tau^4 - (8/3)\tau^3 + \pi\tau^2 \tag{24}
$$

The pdf is given in (25) by taking the derivative of the cdf.

$$f(\tau) = 2\tau^3 - 8\tau^2 + 2\pi\tau \tag{25}$$

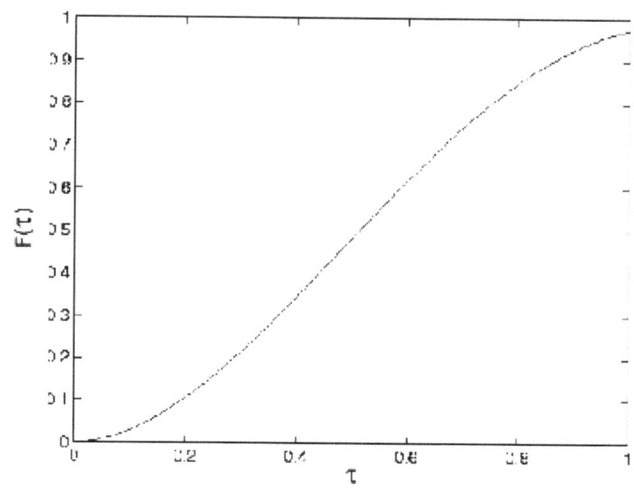

Figure 4. The cdf for the distance between two random points in the unit square.

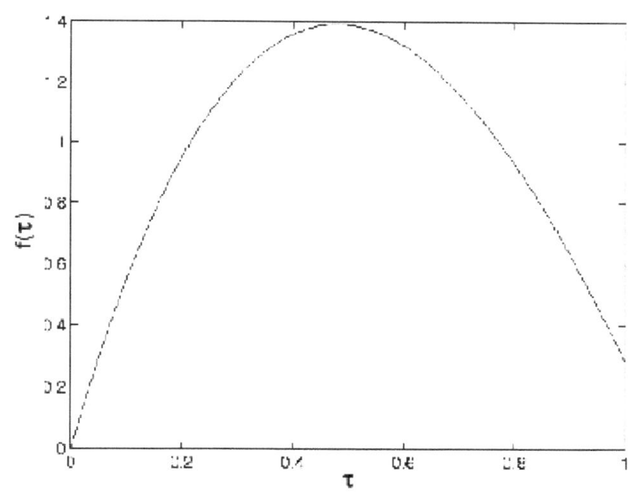

Figure 5. The pdf for the distance between to random points in the unit square.

2. Probability for Two Hop Connection

Consider three nodes uniformly and randomly distributed in the unit square, see Fig. 6.

$$X, Y, Z \in U(K),\ |K| = 1$$

What is the probability that the distance X to Y and Y to Z is less than a threshold τ and the distance X to Z is larger than the same threshold τ ? In other words; what is the probability for having a two hop connection between X and Z and at the same time not an one hop connection for a given max hop distance τ ? (26) We use |.| to describe the area covered by a set.

$$\begin{aligned} P &= P\left(d(X,Y) \leq \tau \cap d(Y,Z) \leq \tau \cap d(X,Z) \geq \tau\right) \\ &= E_{X,Z} P(Y \in (B_\tau(X) \cap B_\tau(Z))) I_2(d(X,Z)) \\ &= E_{X,Z} |B_\tau(X) \cap B_\tau(Z) \cap K| I_2(d(X,Z)) \qquad (26) \\ I_2(v) &= \begin{cases} 1, & v \geq \tau \\ 0, & v < \tau \end{cases} \end{aligned}$$

The probability for having a two hop connection and at the same time not an one hop connection is the overlapping between two circles when the distance between the center of these circles is between τ and 2τ , we use P for this probability (26). It is however rather difficult to calculate the exact function for this value. In this paper we will therefore calculate the upper and lower bound for P .

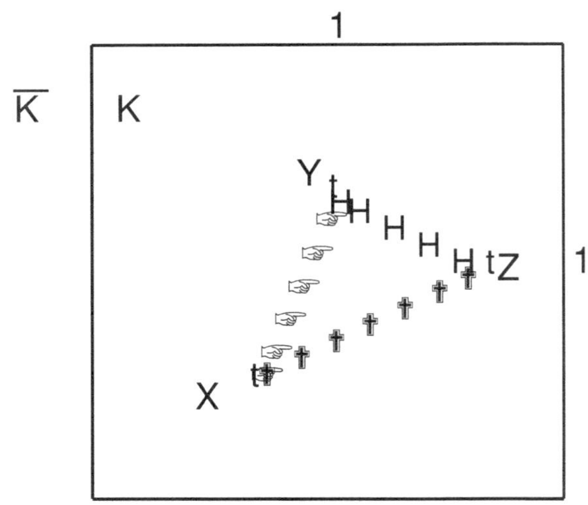

Figure 6. Three points uniformly distributed in the unit square.

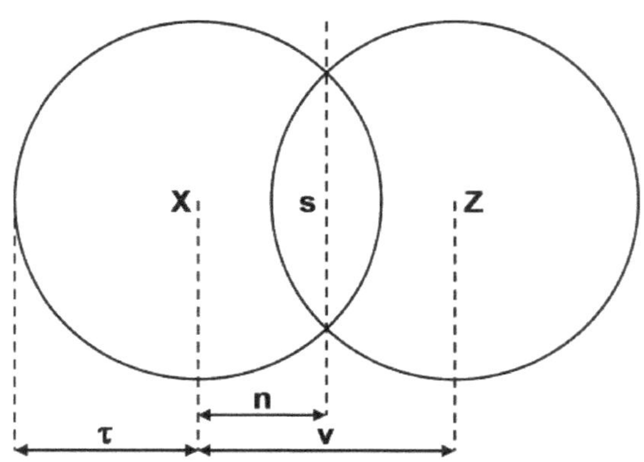

Figure 7. The overlapping area between the coverage of two nodes.

3.1 Upper bound

For the upper bound we use the entire overlapping between the circles, both the part that is inside of the unit square and the part that is outside of the unit square (28). The size of the intersection as a function of the distance between the center of the two circles is given by (27). The expected size of the area is given in (29) and is found by integrating (27) multiplied with the pdf (25) for the distance over the valid distances.

$$B_1(X,Z) = B_\tau(X) \cap B_\tau(Z)$$

$$f_s(v) = 2\left(\tau^2\cos^{-1}\left(\frac{v}{2\tau}\right) - \frac{v}{2}\sqrt{\tau^2 - \left(\frac{v}{2}\right)^2}\right) \tag{27}$$

$$\begin{aligned} P &= E_{X,Z}|B_1(X,Z) \cap K| I_2(d(X,Z)) \\ &\le E_{X,Z}|B_1(X,Z)| I_2(d(X,Z)) = P_u \end{aligned} \tag{28}$$

$$\begin{aligned} P_u &= E_{X,Z} f_s(d(X,Z)) I_2(d(X,Z)) \\ &= \int_{v=\tau}^{2\tau} f_s(v) f(v)\,dv \\ &= \int_{v=\tau}^{2\tau} 2\left(\tau^2\cos^{-1}\left(\frac{v}{2\tau}\right) - \frac{v}{2}\sqrt{\tau^2 - \left(\frac{v}{2}\right)^2}\right) \\ &\quad \cdot \left(2v^3 - 8v^2 + 2\pi v\right) dv \\ &= \left((\pi/3) + ((3\sqrt{3})/4)\right)\tau^6 \\ &\quad + \left(\frac{16\pi}{9} - \frac{36\sqrt{3}}{5}\right)\tau^5 + \left(\frac{3\pi\sqrt{3}}{4}\right)\tau^4 \end{aligned} \tag{29}$$

3.1 Lower bound

The probability for P is given by the total overlapping between the circles (30) minus the part of the overlapping that intersects with the area outside of the unit square (31). To calculate the lower bound for the probability P we will calculate a function, $g_u(\tau)$, that is greater or equal to the part of the overlapping intersecting with the area outside of the unit square (32).

$$f(\tau) = E_{X,Z}|B_1(X,Z)| I_2(d(X,Z)) \tag{30}$$

$$g(\tau) = E_{X,Z}|B_1(X,Z) \cap \overline{K}| I_2(d(X,Z)) \tag{31}$$

$$P = f(\tau) - g(\tau)$$

$$g_u(\tau) \ge g(\tau) \tag{32}$$

$$P_l = f(\tau) - g_u(\tau) \le P$$

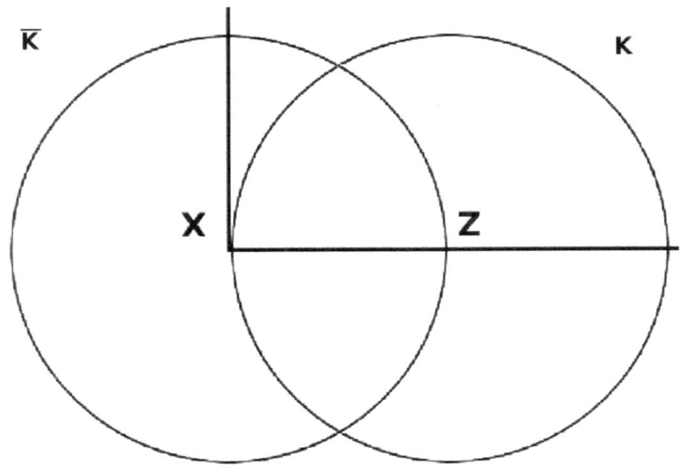

Figure 8. The point X is located in a corner and the point Z is located on an edge.

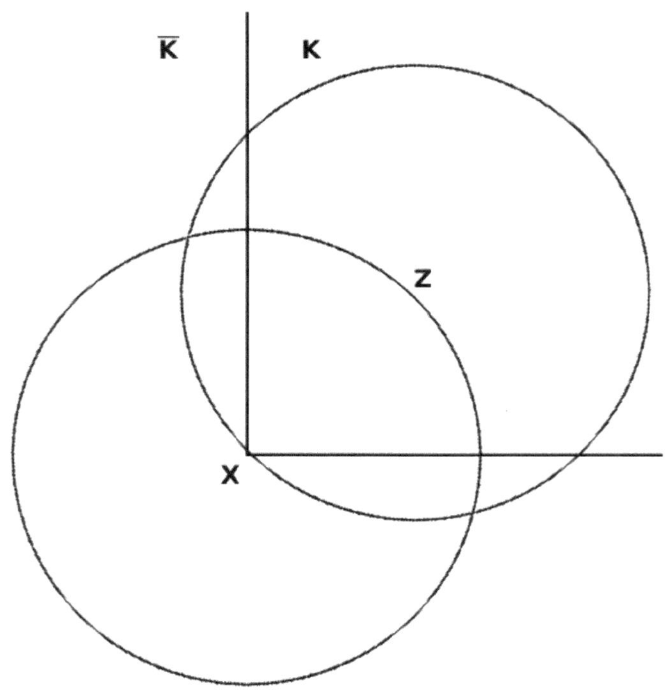

Figure 9. Less than half of the overlapping outside of the unit square.

It is only possible for $B_1(X,Z)$ to intersect with K if $B_\tau(X)$ and $B_\tau(Z)$ also intersects with K. The probability for this is given in (33). For these cases we put the point X at the location giving the largest average value for $B_1(X, Z) \cap K$; this is when X is located in one corner of the unit square. The probability density function for the distance to point Z given that point X is located in the corner is given in (34).

$$
\begin{aligned}
P_o(\tau) &= P((B_\tau(X) \not\subseteq K) \cap (B_\tau(Z) \not\subseteq K)) \\
&= (4\tau(1-\tau))^2 \qquad (33) \\
f_V(v) &= \frac{\pi v}{2} \qquad (34)
\end{aligned}
$$

For all distances between X and Z we set the area outside of the unit square to the highest amount possible. That is half of the total overlapping (35). These two assumptions are based on extensive simulations. The expected size of this area is solved in (36) by using (27), (34), and (33).

$$
\begin{aligned}
g_u(\tau) &= P_o(\tau)\frac{1}{2}E_Z|B_1(x,Z)|I_2(d(X,Z)) \qquad (35) \\
g_u(\tau) &= (4\tau(1-\tau))^2 \qquad (36) \\
&\cdot \int_{v=\tau}^{2\tau} \left(\tau^2\cos^{-1}\left(\frac{v}{2\tau}\right) - \frac{v}{2}\sqrt{\tau^2 - \left(\frac{v}{2}\right)^2}\right) \\
&\cdot \left(\frac{\pi v}{2}\right) dv \\
&= \frac{3\sqrt{3}\pi(1-\tau)^2\tau^6}{2}
\end{aligned}
$$

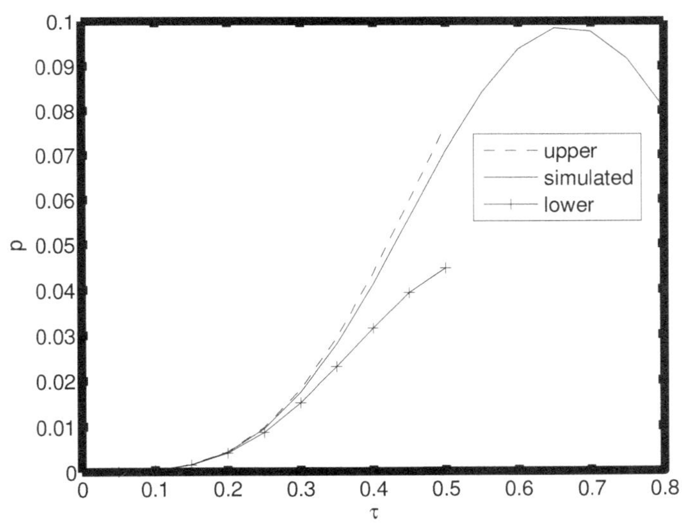

Figure 10. The lower and upper bound compared to the simulated result.

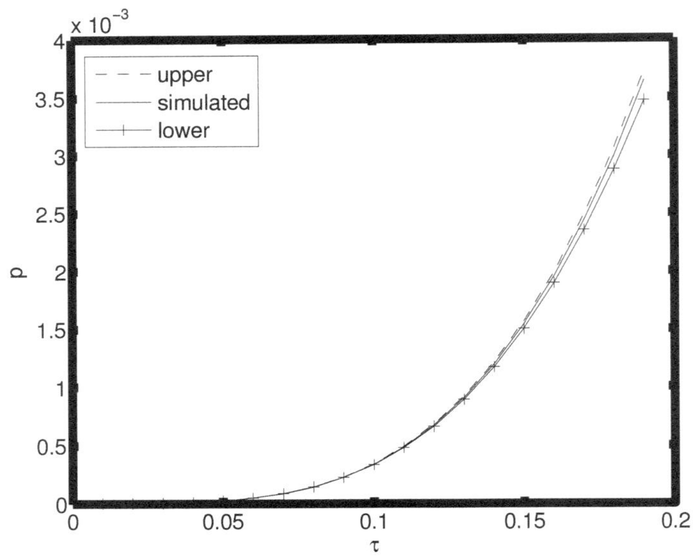

Figure 11. The lower and upper bound compared to the simulated result.

The upper bound for the probability of having a two hop connection between two random points in the unit square is given in (37). The lower bound is given in (38).

$$P_u(\tau) = \left(\frac{\pi}{3} + \frac{3\sqrt{3}}{4}\right)\tau^6 + \left(\frac{16\pi}{9} - \frac{36\sqrt{3}}{5}\right)\tau^5 + \left(\frac{3\pi\sqrt{3}}{4}\right)\tau^4 \quad (37)$$

$$P_l(\tau) = \left(\frac{\pi}{3} + \frac{3\sqrt{3}}{4}\right)\tau^6 + \left(\frac{16\pi}{9} - \frac{36\sqrt{3}}{5}\right)\tau^5 + \left(\frac{3\pi\sqrt{3}}{4}\right)\tau^4 - \frac{3\sqrt{3}\pi(1-\tau)^2\tau^6}{2} \quad (38)$$

4. CONCLUSIONS

In this paper we have presented an exact function for the one-hop connectivity and a lower and upper bound for the two-hop connectivity. These results have been compared to simulated results for the two-hop connectivity. In future work the upper and lower bounds can be improved. It can also be possible to find the exact solution by making use of the joint distribution for the distance between the three points.

References

[1] Paolo Santi and Douglas M. Blough, "The critical transmitting range for connectivity in sparse wireless ad hoc networks," *IEEE Transactions on Mobile Computing*, vol. 2, no. 1, pp. 25–39, 2003.

[2] Feng Xue and P. R. Kumar, "The number of neighbors needed for connectivity of wireless networks," *Wirel. Netw.*, vol. 10, no. 2, pp. 169–181, 2004.

[3] Leonard E. Miller, "Probability of a two-hop connection in a random mobile network," *in Proceedings 35th Conference on Information Sciences and Systems*, Baltimore, March 2001.

[4] Leonard E. Miller, "Distribution of link distances in a wireless network," *Journal of Research of the National Institute of Standards and Technology*, vol. Volume 106, no. Number 2, pp. 401–412, March April 2001.

A NOVEL CLUSTER VALIDATION APPROACH ON PSO-PAC MECHANISM IN AD HOC NETWORK

S.Thirumurugan[1] and E. George Dharma Prakash Raj[2]

[1]Department of Computer Applications, JJ College of Engineering & Technology, Trichirapalli, Tamilnadu, India
s.thiru.gan@gmail.com
[2]Department of Computer Science and Engineering, Bharathidasan University, Trichirapalli, Tamilnadu, India
georgeprakashraj@yahoo.com

ABSTRACT

The ad hoc network places a vital role in contemporary day's communication scenario. This network performance gets up while the clustering phenomenon has been incorporated. The cluster formation using the vital parameters is incredible on deciding the efficiency level of the clustered ad hoc networks. The PSO-PAC mechanism forms clusters based on swarm intelligence by considering energy as crucial parameter. This optimized cluster helps to suits the applications where the energy parameter plays a key role. The clusters formed by this mechanism may not ascertain the compactness of the clusters. Thus, this paper proposes D-PAC as an index based validation mechanism to be handled on clusters formed using PSO-PAC. The cluster formation and validation mechanism have been implemented using OMNET++ simulator.

KEYWORDS

PSO-PAC,D-PAC, Dunn's Index, IPv6.

1. INTRODUCTION

The communication through wireless mode has been an indispensable step in this scenario to suit various applications. This network with ad hoc nature will make the network to be available for various situations. This kind of network comes with the modular growth factor, need to be handled carefully. The clustering gives a solution to this problem. To form the clusters with higher level of efficiency multiple parameters need to be taken into account. It is obviously realized that the distance parameter alone couldn't decide the efficiency of clustering mechanism. Since the nodes which are in wireless network are facing energy drain as a problem with respect to time. Further, the cluster head needs maximum energy among the nodes in a cluster to act as a transceiver. The existing distance based algorithms are showing deficiency in giving an optimum solution. Thus, the multiple parameter based algorithm PSO-PAC(Particle Swarm Optimization of Partitioning Around Clusterhead) has been devised. The swarm intelligence has been the base for the formulation of this efficient clustering mechanism. The behavior of the crow has been considered to devise the procedure of this technique. These optimized clusters need not be compact in their existence. There should be some validation mechanism to analyze the clusters formed. Thus, this study proposes D-PAC(Dunn's Index based Partitioning Around Cluster head) as a cluster validation mechanism to identify the overlapping and separation level of the clusters.

This paper has been organized as follows. Section.1 deals with introduction. Section.2 says about the related works. Section.3 tells about the Ex-PAC mechanism. Section.4 gives out the proposed PSO-PAC procedure. Section.5 deals with the cluster validation mechanism. Section.6 deals with the experimental results. Section.7 brings out the cluster validation results. Section.8 puts down the future works. Section.9 ends up with the concluding remarks.

2. RELATED WORKS

The purpose of clustering has been realized when the protocols like AODV[1] has been integrated with the clustering mechanism[2]. Since this addition to the existing protocols produce an improved result. The single parametric algorithm K-means[3] lacks in identifying an efficient cluster. Thus, PAC[4] procedure has been devised to form an efficient clusters in terms of cluster creation time. But PAC has got the down side on leaving many nodes as outliers. This drawback has made to devise another algorithm on top this PAC. The Ex-PAC[5] came out as an extension to PAC which takes entire nodes and produces the maximum possible clusters. The cluster creation process ultimately improved in Ex-PAC procedure. This approach concludes that Ex-PAC has shown significant improvement over k-means in terms of computational speed. The weighted clustering has faced load balancing and global re-clustering as a challenging issue. Thus, IWCA[6] takes relative mobility of nodes into consideration to improve the weighted clustering.

The EWCA[7] takes cluster set up and cluster maintenance phase into the account. The cluster head leaves the role when the energy of the node goes down below the threshold. This will make some other node to become cluster head for the same cluster. The role of cluster head may be switched among the nodes without waiting for the cluster head to get exhausted in their energy. This may balance the energy[8] utilization of the nodes.

The clustering method needs to minimize the energy to set up cluster, maximize the life time of cluster and ensure the stable structure[9]. The power level of nodes decided the sustainability of the cluster head role. The transmission power alone is not sufficient to calculate the weight of the node. The power reward[10] based weight calculation guarantees the uniform power distribution.

The cluster head detection is based on the signal strength[11]. The nodes are also selected based on signal strength between the cluster head and nodes of the respective cluster.

The dominating sets are recognized in the clustered network. In which the minimum independent set[12] can be constructed and the tree structure of the same can be formed later. The connected dominating set algorithm has been a strong procedure to form the clusters.

The SWCA[13] has been proposed to detect the malicious node which acts as cluster head to tamper the degree computation. This also ensures the secured way of transferring the data to the destination with the help of an authentication mechanism.

The NWCA[14] shows an improved weight based algorithms through challenging parameter calculation methodology for weight. The degree computation has been changed to mean connectivity degree. This novel method also considers the energy level of the nodes to play the role of cluster head.

The DWCA[15] considers the cluster formation based on weight, mobility factor and cluster maintenance. This protocol deals with the node addition to the cluster through distinct approach.

The weight based clustering has been enhanced[16] on the basis of minimizing the load of the cluster head with the help of threshold value. This limits the cluster size to guarantee the cluster head to last longer.

The aforementioned multi parametric algorithms lack in obtaining an efficient cluster. Thus, The multi-parametric algorithm W-PAC[17] has been devised in order to improve the cluster efficiency in terms of time to form the clusters.

The multi-parametric swarm intelligence based clustering mechanism PSO-PAC[18] takes the necessary parameters to identify the cluster head of the cluster. This parameter optimization will suit specific application.

The IPv4 address no longer will serve the world has been realized. The IPv6 address has got the focus in the present scenario. The configuration of address can be done in two ways. The configuration using DHCPv6 server and stateless autoconfiguration[19]. This configured address can be local link address or global address. This will be decided based on the application requirement. But there must be awareness on the limitations of using the address range has been vital.

The stateless autoconfiguration of IPv6 nodes should have duplicate address detection mechanism. Since when the nodes move across the clusters there is a chance of address duplication. To eliminate this passive duplicate address detection mechanism[20] has been introduced. This method shows better results than passive autoconfiguration for mobile ad hoc networks.

The purpose of autoconfigured address has been recognized while the ad hoc network has come to reality. The limitation of IPv4 and the need of IPv6[21][22] has been understood clearly. Having understood the difference between IPv4 and IPv6, the scenarios will demand the specific way of addressing the nodes. These works are dealing with the autoconfiguration or manual configuration of IPv4 and IPv6 nodes. The clustering as a mechanism comes for this IPv4 or IPv6 autoconfigured node to make the routing simple and also confirm the efficient utilization of bandwidth and resources.

The cluster formation algorithms devised until now missed out on appropriate re-clustering phenomenon. The re-clustering should be based on identifying the strength of the existing clusters. The role of various indices[23] on evaluating the cluster should be understood very well. The cluster classification[24] also plays key role in determining the perfectness of the clusters. Those classifications are of numeric, discrete and partitioned types. It also finds out the preferred clustering method for a given sample set of nodes.

3. EX-PAC

The PAC creates the clusters based on Manhattan distance. The Manhattan distance saves time in computing the distance between pair of nodes. The results achieved are not adequate to find out the appropriate cluster in the ad hoc scenario network. This has been further enhanced through Ex-PAC algorithm which has been laid on top of PAC. The experimental results show that Ex-PAC has given better results than K-means algorithm in forming clusters. The formula (1) puts down the calculation of Manhattan distance.

$$\text{Manhattan Distance} = \sum_{i=1}^{n} |xi - yi| \qquad (1)$$

3.1 Ex-PAC algorithm

1) Assume Ni = temporary Cluster head.
2) Compute Manhattan distance between pair of nodes.
3) if (MD < range)
   Begin
      Add(Ni,Ci)
      Count = Count + 1
   End
4) Repeat the steps 1 through 3 till all the nodes in the cluster are examined.
5) Select the First Cluster which has maximum count value.
6) Select the non cluster nodes.
7) Choose the cluster which includes non cluster nodes.
8) Select cluster returned in step 7 as Second Cluster.
9) Repeat the steps 6,7 and 8 until there is no change in Cluster formation.

The Ex-PAC procedure has produced remarkable improvement over the PAC procedure. But still lacks in identifying the perfect cluster head of cluster. The cluster head couldn't be decided based on distance factor only. There should be some mechanism over the Ex-PAC.

4. PSO-PAC

The crow behaves noticeably when it finds foodstuff. It alerts neighbours after seeing the eatables through the acoustic signal. Thus, a group of crows will create the cluster. This formation is based on more than one parameter. Those parameters are distance and energy of the crows in communication process. This multi parameter specifies that the cluster formation with the help of swarm intelligence makes the clustering process in a highly efficient way. This behaviour of the crow can be considered in making dynamic clusters.

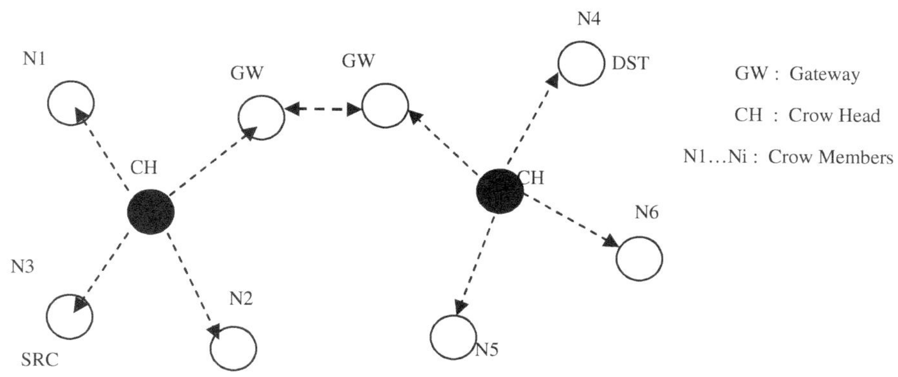

Figure 1. Network Model :Two Clusters

The crow has been taken as a particle in this cluster formation. This particle plays the role of cluster head and members as well. The crow head is the particle which plays cluster head role and crow member is the node belongs to the cluster. The eatables have been considered as a destination to be reached and energy of the crow will be the source of the particle. The activity here is the particle makes the communication after seeing the eatables. Similarly in network communication the source sends the data after knowing the destination through cluster head node. The Fig.1 shows the network model consist of two clusters where each cluster has separate crow head node. The gateway acts as mediator between two clusters. This ensures inter-cluster communication to happen. In this model the destination is far away from the source. This network setup will be retained till the communication between the source and destination gets over. After the data transfer gets over then network has to be reformed with new energy source and destination. In this way the cluster formation happens completely dynamic in nature which suits very well with the behaviour of the crow.

4.1 Cluster Formation by Crow Behaviour

1) Input the clusters formed in Ex-PAC procedure.
2) j = j + 1
3) CH = Max_Energy_Node(Cj)
4) Add to Newcluster(Kj,CH)
5) Cluster = Cj
6) i = i + 1
7) E= Get_energy(Ni,Cluster)
8) If (E < EnergyThreshold)
   Newcluster (Kj, Ni)
9) Repeat the steps 4,5 & 6 until all the nodes belong to the cluster Cj gets examined.

Repeat the steps 2 through 7 until all the clusters of network are examined.

The Max_Energy_Node function returns the node with maximum energy for each cluster. This will be considered as cluster head of the cluster.

4.1.1 IPv6 configuration of Clusters

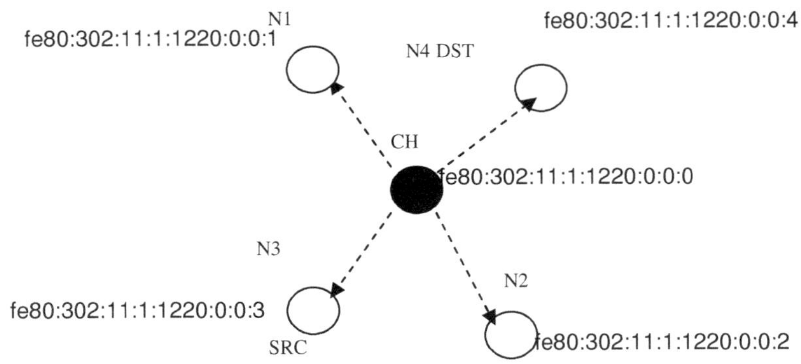

Figure 2. IPv6 Configured Network Model

The cluster formation takes the output of Ex-PAC procedure and recreates the cluster dynamically based on the energy level of nodes. The Fig.2 shows the IPv6 nodes to form the clusters dynamically in a stateful way.

4.1.2 IPv6 Configuration Procedure

1) Input the clusters from crow behaviour procedure.
2) Install the DHCPv6 server process in Cluster head node.
3) Cluster head sends hello message to all the members.
4) For each Cluster Ci
   Begin
     For each member node Nj
       Begin
         Node Nj sends reply to the Cluster head.
         Cluster head sends an IPv6 address to Nj.
       End
   End
5) Repeat the step 4 for j = 0 ... Number of Nodes in cluster Ci and for i = 0 to Number of clusters.

The IPv6 configuration procedure configures the nodes belong to the clusters under stateful way using DHCPv6 server. In this method the cluster head sends hello message to all the members at the initial stage. The member nodes send the reply to the cluster head. This reply informs the member belong to the cluster where the cluster head exists. The cluster head now sends IPv6 address to the member node as a part of configuration.

5. D-PAC : CLUSTER VALIDATION

This parameter helps to validate the clusters formed using PSO-PAC procedure. This validation process will identify the clusters compactness and the separation of the clusters based on the parameter value computed. The formula (2) shows that the min distance between the clusters Cp and Cq. This could be obtained by taking all the nodes belong to the two clusters to apply the distance computation such as Manhattan distance.

$$\mathrm{dis}(Cp, Cq) = \min_{m \in Cp,\, n \in Cq} \left(d(m, n) \right) \tag{2}$$

In the formula (3) the max diameter of the cluster will be identified with the help of the pair of nodes which are placed far apart within the cluster. This process takes all the nodes within the cluster.

$$\mathrm{dia}(Cp) = \max_{x,y \in Cp}\left(d(x,y)\right) \qquad (3)$$

The formula (4) gives out the dunn's index value which is ratio between the minimum distance of clusters and maximum diameter of cluster.

$$\mathrm{Min}_{\,p=1..nc}\left\{\mathrm{Min}_{\,q=i+1..nc}\left[\frac{\mathrm{dis}(Cp,Cq)}{\mathrm{Max}_{r=1...nc}\,\mathrm{dia}(Cr)}\right]\right\} \qquad (4)$$

The dunn's index value calculated using the above mentioned formula tells how the clusters are segregated with each other and also represents the strength of the cluster based on the nodes exist.

5.1 D-PAC Algorithm

1. Create the clusters using the PSO-PAC procedure.
2. Find the distance between pair of clusters Cp and Cq.

$$d(m,n) = \sum_{i=1}^{N} |mi - ni|$$

Dist(Cp,Cq) = min(d(m,n)) , m∈Cp , n∈Cq

3. Repeat the step 2 for all the pairs of clusters.
4. Find the maximum diameter of the cluster.

Diameter(C) = max(d(m,n)) , m&n∈C

5. Repeat the step 4 for 1 to N clusters.
6. Calculate Dunn's Index.

Dunn's index = Min(dist(Cp,Cq)) / Max(diameter(Cr))

7. if Dunn's index between 0 and 0.5 then
   Clusters are compact and separation is less.
   Else
   if Dunn's index between 0.51 and 1.0
   Clusters are compact and well separated.

This algorithm takes PSO-PAC output as input clusters for the validation process. The Dunn's index has been calculated based on the formulas aforementioned. The resultant value will reveal that the clusters are overlapping or well separated.

6. EXPERIMENTAL RESULTS

This PSO-PAC algorithm and D-PAC validation mechanism has been implemented in OMNET++ and the results are tabulated. This work has been carried out with the system configuration of 64bit AMD processor, 2GB RAM and windows XP as an operation system. This simulation has been done for 25 nodes and 50 nodes.

Table 1. Simulation Parameters

Parameter	Values
N (Number of Nodes)	25 , 50, 300
Space (area)	100 ×100
Tr (Transmission range)	20m
Execution Time	5 sec
Threshold(Energy) in Units	500

The Table.1 shows the simulation parameters considered while simulating the study.

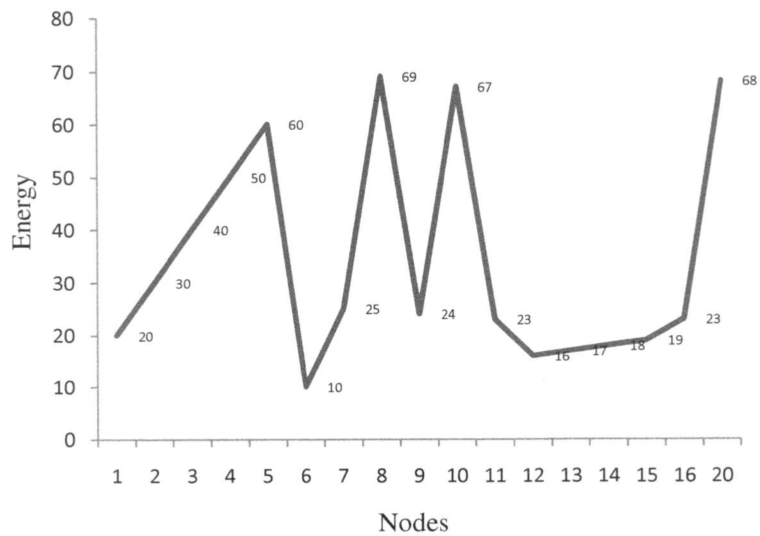

Figure 3. Energy Graph for 25 Nodes : Cluster C1

The cluster heads could be fixed and need to be changed in accordance with the time T. The energy level graph Fig.3 at a specific time T identifies the node which is eligible to be considered as cluster head at the time period T1. After time period T1 some other node will be chosen as cluster head at time period T2. This graph clearly reveals the cluster heads selection in the order.

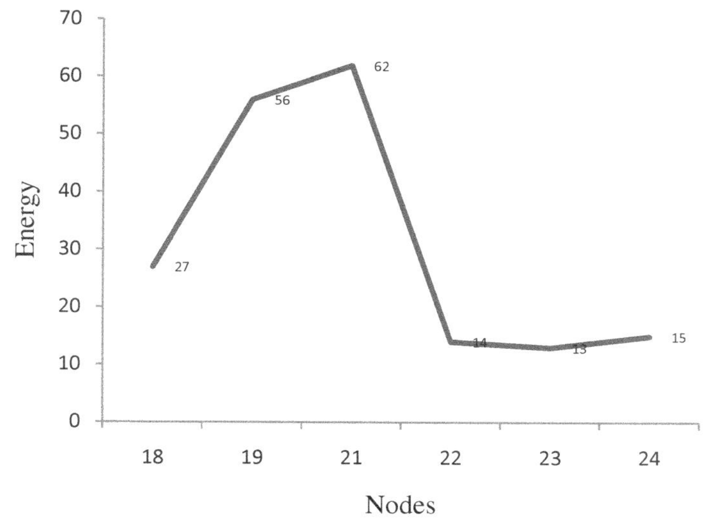

Figure 4. Energy Graph for 25 Nodes : Cluster C2

The Fig.4 shows the nodes and their respective energy level belongs to cluster C2. This ushers that the node21 will be chosen as cluster head since it has high energy level at time T1.

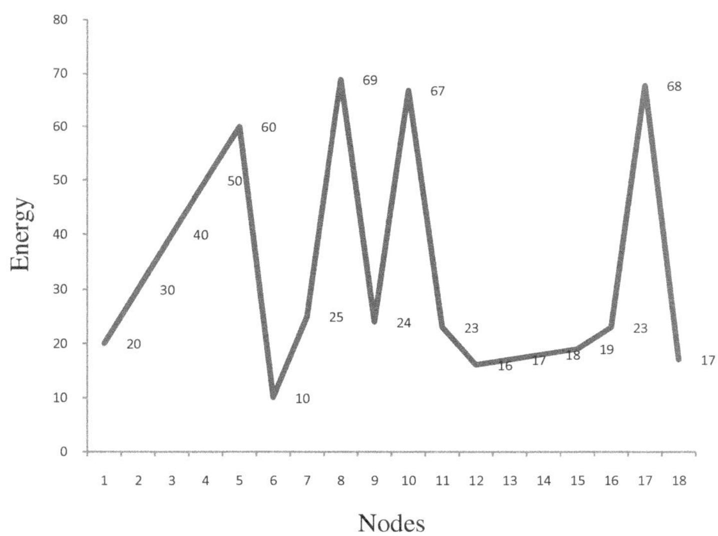

Figure 5. Energy Results for 50 Nodes : Cluster C1

The Fig.5 ushers the energy based results for the sample size of 50 nodes. This shows that node8 takes the role of cluster head.

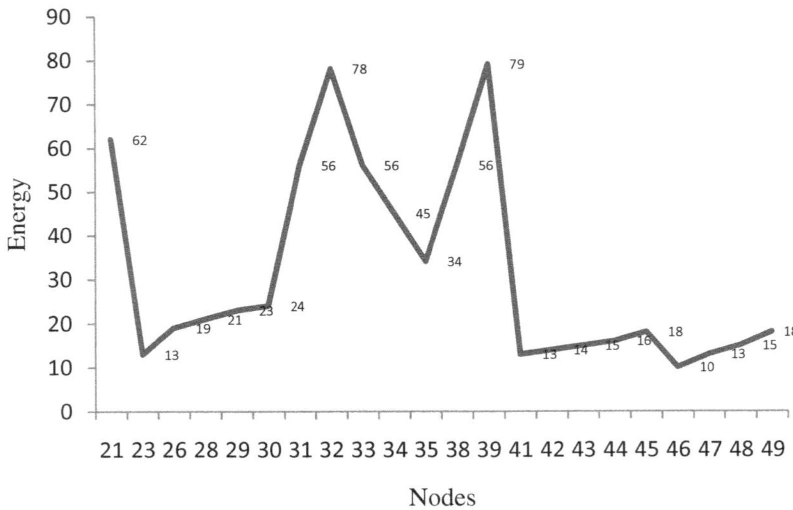

Figure 6. Energy Results for 50 Nodes : Cluster C2

The Fig.6 ushers the energy based results for the cluster2 where the sample size has been 50 nodes. This identifies node39 as the cluster head.

7. D-PAC : Cluster Validation Results

The clusters formed using PSO-PAC will not ascertain the maintenance of the clusters. There should be some mechanism to analyze the clusters formed to confirm the compactness of the clusters. This tells the clusters separation and overlapping percentage.

Table 2. Results : D-PAC

Num. Of Nodes	Dunn's Index	Separation	Overlap	Compact
25	0.52	52%	48%	High
50	0.48	48%	52%	Low
300	0.01	10%	90%	Very Low

The Table.2 shows the results obtained in a tabulated manner. It is clearly reveals that the value of index indicates the segregation of clusters and also the level of cluster overlap. The calculation gives 0.52 as the Dunn's index result for 25 nodes. This value indicates the separation of clusters. The clusters are 52% apart has been identified through this calculation. This drives to understand that there is a chance of 48% overlap in clusters of network. The index value for 50 nodes 0.48 has been obtained says that the clusters separation is less than that of 25 nodes. This proportionately increases the overlapping of the clusters which is slightly higher when compared with 25 nodes. The compactness is decided based on the intra cluster relationship of nodes. The separation is high which indicates the high intra cluster relationship ends up at high compactness. The compactness for 50 nodes has been low when compared with the 25 nodes. The compactness for 300 nodes has been very less to indicate the closeness of clusters. This will increase the interference between the clusters. This can be solved by invoking re-clustering process.

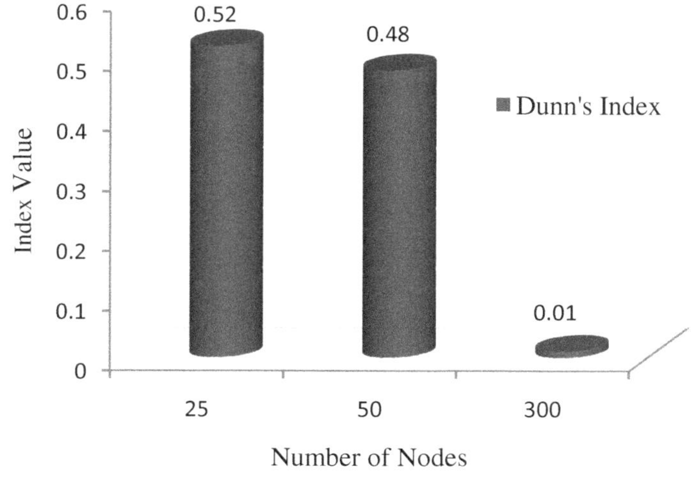

Fig.7 Graphical illustration

The Fig.7 shows the graph of Nodes against Dunn's index. This illustration makes to understand the index value for the different sample size. The higher the value of index the stronger the cluster and separation would be high.

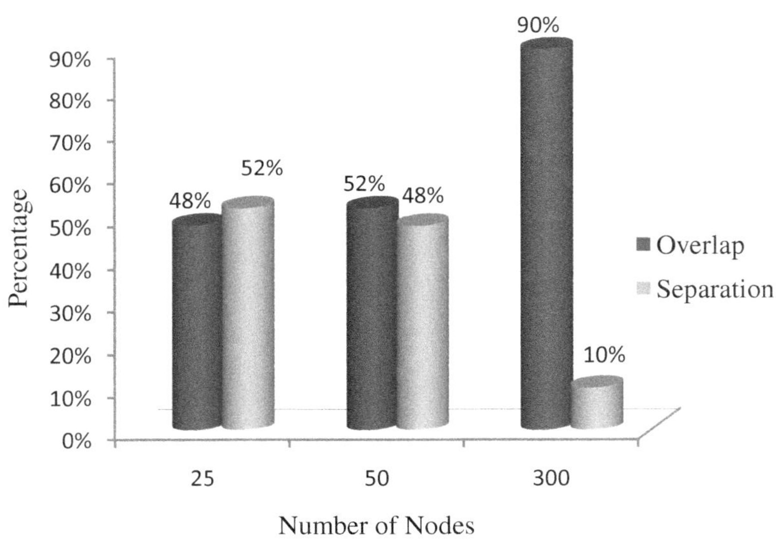

Fig.8 Graphical illustration

The Fig.8 tells the possibility of overlapping based on the separation which in term computed based on the Dunn's index. This graph makes to visualize the occurrence of overlapping clusters in the selected sample set of nodes.

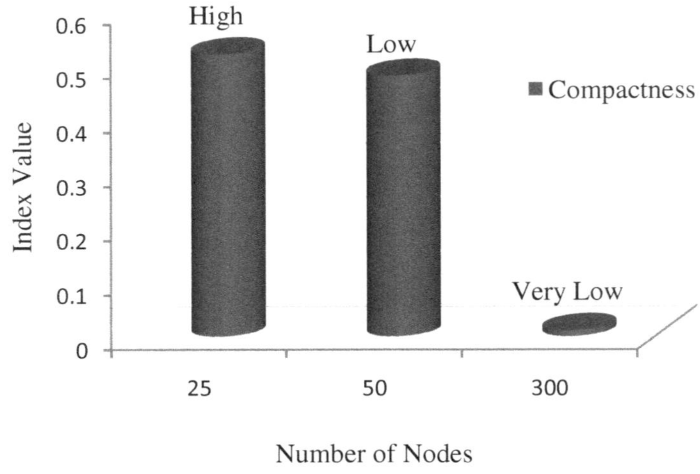

Fig.9 Graphical illustration

The Fig.9 identifies the compactness level of the clusters. This shows that the smaller nodes sample has high compactness than larger one.

8. FUTURE DIRECTIONS

This work has considered dunn's index as a validation parameter on clusters. This work should be done for various validation parameters to identify the strength of clusters.

9. CONCLUSION

This study has devised the clusters using PSO-PAC algorithm. These clusters are validated for their compactness using D-PAC procedure. This D-PAC result obviously determines the closeness of the clusters. This closeness helps to decide about the clusters which are interfering due to overlapping. In this approach intra-cluster strength has been measured by separation percentage. The higher separation level makes the understanding that the bandwidth could be reused with less or without interference. The periodical validation of these measures tells the time of re-clustering the network. The implementation using OMNET++ confirms the clusters separation level with the help of the Dunn's index value.

REFERENCES

[1] S.Thirumurugan, (2010) "Direct sequenced C-IAODV Routing Protocol", International Journal of computer science and Technology, vol.1(2), pp108-113.

[2] S.Thirumurugan. (2011) "C-AODV: Routing Protocol for Tunnel's Network", International Journal of computer science and technology, vol.2(1), pp113-116.

[3] Jieming Wu, Wenhu Yu, "Optimization and improvement based on K-Means Cluster algorithm", Second International Symposium on Knowledge Acquisition and Modeling, 2009,pp.335-339.

[4] S.Thirumurugan, E.George Dharma Prakash Raj, (2011) "PAC - A Novel approach For Clustering Mechanism in Adhoc Network", ICSCCN'11, pp 593-598.

[5] S.Thirumurugan, E.George Dharma Prakash Raj, (2012) "Ex-PAC : An Improved Clustering Technique in Ad Hoc network", RACSS'12, pp195-199.

[6] Likun Zou, Qishan Zhang, Jianwei Liu, "An Improved Weight-Based Clustering Algorithm in MANETs", WiCom'08, 2008, pp.1-4.

[7] Chang Li, Yafeng Wang, Fan Huang and Dacheng Yang, " A Novel Enhanced Weighted Clustering Algorithm for Mobile Networks",WiCOM'09, 2009, pp.1-4.

[8] Zhang Lei ,Wang Xuehui, " An energy balanced clustering algorithm for mobile ad hoc networks",WiCOM'09, 2009, pp.1-4.

[9] Jing An, Chang li, Bin li, "A Improved weight based clustering algorithm in mobile ad hoc networks", YC-ICT'09, 2009, pp.220-223.

[10] Na Zhang, Ningqing LIU, Weixiao MENG, Qiyue YU, " A Novel Weighted Clustering Algorithm Based on Power Distribution for Cooperative Communication Network", PCSPA'10, 2010, pp.142-145.

[11] Dao-QuanLI, Wei-HuaXUE, Qi-Guang CAO, Huai-Cai WANG, "Research on Ad Hoc network clustering algorithm based on signal strength", ICIS'10, 2010, pp.719-722.

[12] Dao-QuanLI, Wei-HuaXUE, Qi-Guang CAO, Huai-Cai WANG , "A New Distributed CDS Algorithm of Ad Hoc Network Based on weight", ISDEA'10, 2010, pp. 90-94.

[13] Yingpei Zeng, Jiannong Cao, Shanqing Guo, Kai Yang, Li Xie, " SWCA: A Secure Weighted Clustering Algorithm in Wireless Ad Hoc Networks",WCNC'09, 2009, pp.1-6.

[14] Yang Wei-dong, " Weight-Based Clustering Algorithm for Mobile Ad Hoc Network", CSQRWC'11, 2011, pp.787-791.

[15] Naveen Chauhan, Lalit Kumar Awasthi, Narottam Chand, Vivek Katiyar and Ankit Chugh, " A Distributed Weighted Cluster Based Routing Protocol for MANETs. International Journal of Wireless Sensor Network", Vol.3(2), 2011, pp.54-60.

[16] Sudhakar Pandey, Narendra Kumar Shukla, " Improved Weighted Clustering Algorithm for Mobile Ad Hoc Networks", IJEMS,Vol 2(1), 2011, pp.20-25.

[17] S.Thirumurugan, E.George Dharma Prakash Raj, (2012) "W-PAC: An Efficient weighted Partitioning Around Cluster Head Mechanism for ad hoc network", CCSEIT'12, pp182-188.

[18] S.Thirumurugan, " PSO-PAC: An Intelligence Clustering Mechanism in Ad hoc Network",NETCOM'12, pp.55-62.

[19] LiQing, "An IP address auto-configuration scheme for MANET with global connectivity", ICIME'10, 2010, pp.244-247.

[20] M Sivakumar , C J Jickson , "R M S Parvathi ,Passive Duplicate Address Detection in DYMO Routing Protocol for MANETS",ICVCI'11, 2011, pp.9-14.

[21] He Zhonglin ,He Yuhua, "Study on Key Technologies of MANET", ISME'10, 2010, pp.112-115.

[22] Ulrich Herberg, Thomas Clausen, "Yet Another Autoconf Proposal (YAAP)for Mobile Ad hoc NETworks", MASN'10, 2010, pp.20-26.

[23] Mahesh Visvanathan Adagarla, B Srinivas Gerald, H Lushington Peter Smith, (2009) "Cluster validation: An integrative method for cluster analysis", IEEE International Conference on Bioinformatics and Biomedicine workshop, pp238.

[24] Rama,B., Jayashree,P., Salim Jiwani, (2010) "*A Survey on clustering current status and challenging issues*, International Journal on computer science and engineering, 2(9), pp2976-2980.

FRAME AGGREGATION MECHANISM FOR HIGH-THROUGHPUT 802.11N WLANS

Jyoti Kolap#, Shoba Krishnan*, Ninad Shaha$

#Department of Electronics and Telecommunication Mumbai University.

#kolapjyoti@gmail.com

* Department of Electronics and Telecommunication Mumbai University.

*shobakrishnan@hotmail.com

$ Computer Centre, IIT Bombay.

$ninadshaha@iitb.ac.in

ABSTRACT

IEEE 802.11n Draft is next-generation wireless LAN standard. This article gives detailed description and comparative study of A-MPDU and Two level frame aggregation mechanisms proposed in the latest 802.11n draft standard, which improve extensively the channel efficiency and data throughput. Ns 2 simulator is used for simulation of 802.11n WLAN frame aggregation. Simulation results confirm that A-MPDU, and Two level frame aggregation methods improve extensively data throughput. It analyse the performance of each frame aggregation scheme in distinct scenarios, and it conclude that overall, the two-level aggregation can improve throughput performance of 802.11n. Two level aggregation can more effectively deliver data from multiple sources, PDR rate is much higher with much lower delay and Two level aggregation consumed less energy as compared with A-MPDU approach..

KEYWORDS

NS-2 IEEE 802.11N, Frame Aggregation, A-MPDU, Two level aggregation,

1. INTRODUCTION

In the recent years, wireless LAN (WLAN) is most popular for mobile equipment's. Because the equipment cost decreases and the data delivery rate becomes higher. IEEE802.11 WLAN is most popular network in wireless networking, because of its advantages its system possesses such as interoperability, mobility, flexibility, and cost-effective deployment. In July 2002, to achieve higher data rate, IEEE 802.11 standard working group established the High-Throughput Study Group (HTSG). The IEEE 802.11n ("n" represents next- generation) resolution to compose a high-throughput (HT) extension of the current WLAN standard that will increase transmission efficiency, throughput and reduce compulsory overhead. The 802.11n standard offers several advantages over previous wireless LAN technologies.

The most majorable advantages are substantially improved reliability and greater application data throughput. However, the low transmission rate of its medium access control (MAC) and a physical (PHY) layer protocol restrict to support high data rate application. Current WLAN systems having difficulties with the increasing expectations of users and with volatile bandwidth and delay-boundary demands from new higher data rate services, such as high-definition television (HDTV)), file transfer, and online gaming, video teleconferencing, multimedia streaming, voice over IP (VoIP. The major aim of the IEEE 802.11n Task Group (TGn) is to have maximum data throughput of at least 100 Mb/s. IEEE 802.11n gives recent propositions for PHY which includes multiple input multiple output (MIMO) antennas with orthogonal frequency division multiplexing (OFDM) and various channel binding mechanism.

However, it has been investigated that simply improves the PHY data rate, and it will not be sufficient for enhancing the system throughput. TGn proposed mac enhancement method in latest IEEE 802.11n standard which enhance throughput and improves efficiency. One of the problems with the legacy MAC is the overhead associated with MAC header.[2] One of the efficient ways to reduce the overhead caused by MAC header is to use larger packets. So, aggregating a number of frames together for transmission is becoming popular for high efficient MAC [1].

The throughput performance at the MAC layer can be improved by enhancing mac layer. It can be done by combining several frames before transmission and i.e. frame aggregation. Frame aggregation not only reduces the transmission time for compulsory overheads, but also reduces the waiting time during CSMA/CA (Carrier Sense Multiple Access – Collision avoidance) random backoff period for successive frame transmissions.

This article proposed a highly efficient and flexible frame aggregation scheme, which can substantially improve the channel efficiency. The proposed frame aggregation can occur at two levels, namely the A- MSDU (MAC Service Data Unit) level and A-MPDU Aggregate-MAC Protocol Data Unit level. Besides, the aggregation at each level can be used together or separately.

Table1. 802.11 STANDARDS and specifications

802.11	**Freq. (GHz)**	**Throughput (Mb/s)**	**Max bit rate (Mb/s)**	**Modulation**	**$r_{in.}$ (m)**
a	5	23	054	OFDM	~35
b	2.4	04.3	011	DSSS	~38
g	2.4	19	054	OFDM	~38
n	2.4, 5	74	248	OFDM	~70

Hence, this two level aggregation can operate flexibly, based upon the applications and real traffic scenarios. The main aim of this article is to implement the new approach, Two level aggregation for more improved performance and to analyse the performance of each frame aggregation scheme in distinct scenarios, to analyse end to end delay, energy consumption and packet delivery ratio and compare the performance of draft of the IEEE 802.11n A-MPDU aggregation schemes along with the two-level aggregation.

2. IEEE 802.11N

IEEE 802.11 is a set of standards for wireless local area network (WLAN) computer communication, developed by the IEEE LAN/MAN Standards Committee (IEEE 802) in the 5 GHz and 2.4 GHz public spectrum bands. Although the terms 802.11 and Wi-Fi are often used interchangeably, the Wi-Fi Alliance uses the term "Wi-Fi" to define a slightly different set of overlapping standards. The 802.11 family includes over-the-air modulation techniques that use the same basic protocol. The most popular are those defined by the 802.11b and 802.11g protocols, and are amendments to the original standard. Some of the important protocols are

IEEE 802.11n is a proposed amendment to the IEEE 802.11-2007 wireless networking standard to significantly improve network throughput over previous standards, such as 802.11b and 802.11g, with a significant increase in raw (PHY)[1] data rate from 54 Mb/s to a maximum of 600 Mb/s. It is also known as **draft-n.**The 802.11n standard will offer several advantages over previous wireless LAN technologies. The most majorable advantages are substantially improved efficincy and greater application data throughput. However, the low efficiency of its medium access control (MAC)[1] and physical (PHY) layer protocols restricts its applications to support high data rate multimedia services.

3 MAC Layer Overhead Reduction

The PHY enhancements described thus far increase maximum PHY data rate, but would make very inefficient use of the airwaves without the following MAC layer enhancements also utilized by 802.11n.

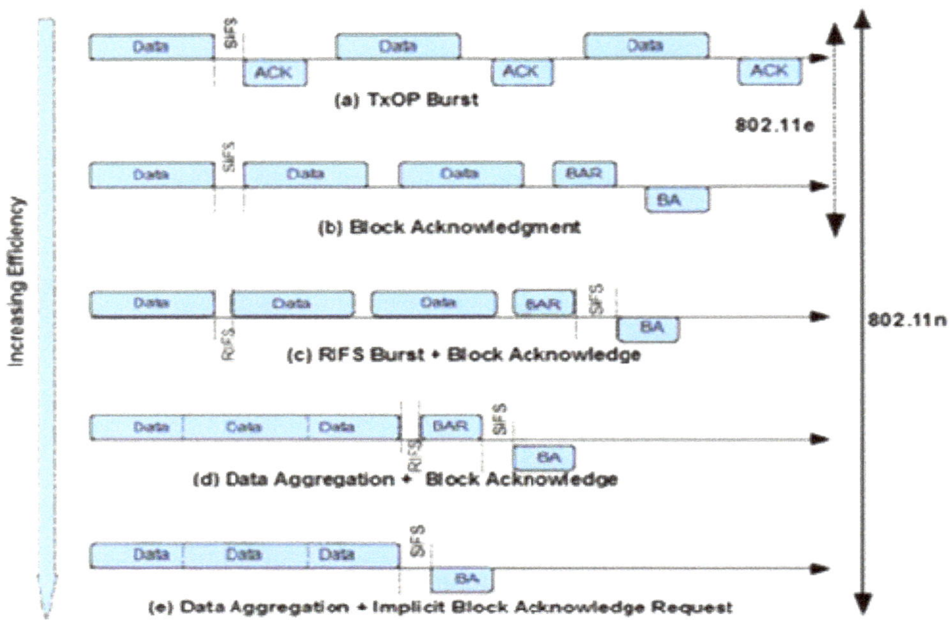

Fig No.1 Block Acknowledgement

3.1. Block Acknowledgement -

Block Acknowledgement reduces the number of ACKs that a receiver must send to a transmitter to confirm frame delivery. Legacy 802.11a/g transmitters expect an (almost immediate) ACK for each non-multicast/broadcast frame. But 802.11n transmitters also accept Block ACKs which confirm receipt of multiple unicast frames. For eg, instead of sending 9 legacy ACKs to confirm frames 1 through 8 and 10, an 802.11n receiver can say the same thing with just one Block ACK.

3.2. Frame Aggregation Mechanism

Aside from the RF improvements, 802.11n has also improved the efficiency of the MAC (Media Access Control) protocol using a single ACK mechanism for multiple frames and ability to aggregate multiple frames into a single transmission.

In 802.11abg, each frame is acknowledged immediately following each transmission. This added robustness comes at a price – no one can transmit a frame while the current sender is waiting for an ACK.802.11n adds a block ACK (BA) mechanism in which multiple frames can be streamed out and acknowledged by a single frame. This cuts the wait time between frames transmission and allows just the missing frames or frames received in error to be resent by checking a compressed bit map. The transmitter need send only one BA request (BAR) to the receiver. If the receiver acknowledges it, the BA mechanism may be used for the duration of the physical connection (such as between a client and access point). By combining frames together into one larger frame, 802.11n can send multiple frames per single access to the channel

The draft 802.11n specification was crafted with the previous standards with compatibility with more than 200 million Wi-Fi devices which currently in use. An 802.11n access point will communicate with 802.11a devices on the 5-GHz band as well as 802.11b and 802.11g hardware on the 2.4-GHz frequencies. In addition to basic interoperability between devices, 802.11n provides greater network efficiency over what 802.11g offers in mixed mode.

Network efficiency is basically the proportion of the available bandwidth that is used to transmit data as opposed to overhead or protocols used to manage network communications. Wireless environments are much more challenging to orchestrate than wired networks, so there is generally more overhead to ensure that data sent is actually received, and that other clients leave the channel open during transmission. The presence of 802.11b nodes makes communications difficult on the 2.4G-Hz band because the older standard does not recognize OFDM, which is employed by802.11g and draft-n. This means that if OFDM clients want to communicate in the presence of 802.11b clients, they need to use the older standard's communication protocol at least to protect their higher-rate OFDM transmissions. This drops network efficiency considerably because data packets take far less time to transmit with 802.11g and draft-n than they do under the old 802.11b standard. Some WLAN suppliers, including Broadcom, devised innovative schemes to improve the efficiency and data throughput of mixed 802.11b/g networks.

One of the most important features in the draft-n specification is frame aggregation to improve mixed mode performance. Rather than sending a single data frame, the transmitting end user bundles several frames together. Thus, aggregation improves efficiency by restoring the percentage of time that data is being transmitted over the network,

The main medium access controller (MAC) feature that provides a performance improvement is aggregation. The types of aggregation are defined:

1. Aggregation of MAC Service Data Units (MSDUs) at the top of the MAC (referred to as MSDU aggregation or A-MSDU)

2. Aggregation of MAC Protocol Data Units (MPDUs) at the bottom of the MAC (referred to as MPDU aggregation or A-MPDU)

3. Two level aggregation

3.2.2. A-MSDU Mechanism

In the Aggregated Mac service data units i.e. A-MSDU scheme, multiple MSDUs are bundled to form a MPDU which could consist of multiple sub frames either from multiple sources or for multiple destinations. Fig.2 illustrates the frame format of an A-MSDU. An A-MSDU consists of multiple sub frames (i.e. multiple MSDUs). Each sub frame of an AMSDU has a sub header (Destination address, Source Address, (Length), MSDU, and padding bytes. The size of the MSDU in each sub frame can be different. Different size of MSDUs in each sub frames are aggregated. To make the length of the sub frame in multiple of 4 bytes except for the last sub frame the padding bytes are appended. All the sub frames are bundled and share a common MAC header and frame check sequence (FCS) which is calculated over all the sub frames and a common MAC header and then appended as the trailer.

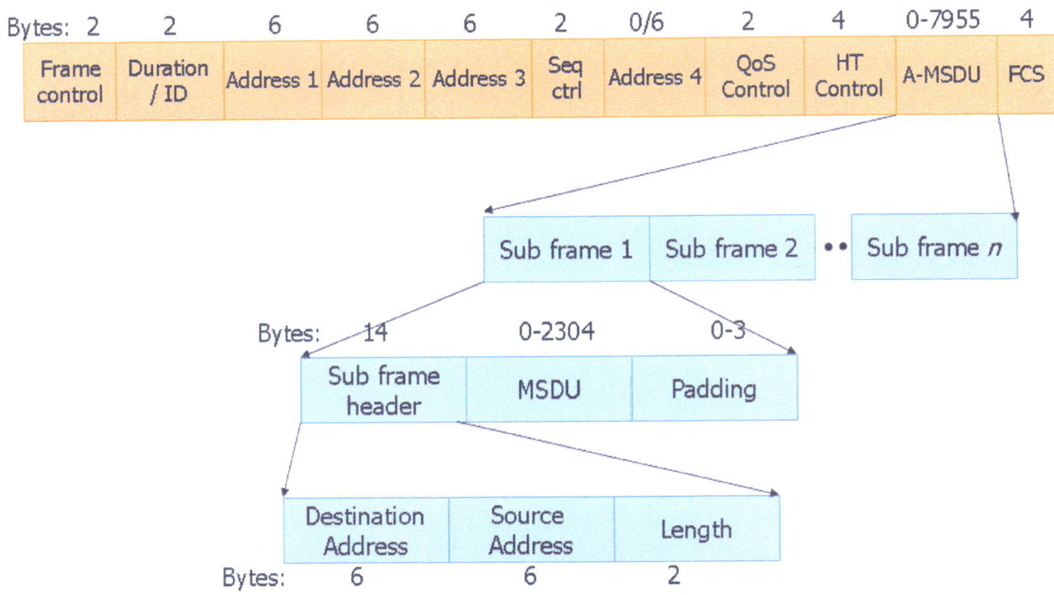

Fig.No. 2 frame format of A-MSDU

By the PHY layer the A-MSDU frame is considered as a single MPDU. Since there is no frame checksum for the individual sub frames, selective retransmission of corrupted sub frames is not possible. Sub frames have the same sequence number and traffic identifier (TID). The maximum length of an A-MSDU frame can be 3839 or 7955 bytes. This capability information is exchanged during the time of association.

The common MAC header contains only source and the access point's (AP's) address in uplink transmissions.Only in the sub headers, the corresponding destination addresses are present., a station can pack MSDUs destined for multiple destinations and send it to the AP in the uplink transmission., the AP can combine multiple MSDUs from various source addresses and send it to a single destination in downlink transmission. An A-MSDU frame cannot be transmitted to multiple end receivers.

3.2.3. A-MPDU Mechanism

Multiple MPDUs with a common PHY header are packed as an A-MPDU which can contain several MSDUs and/or A-MSDUs. The size of each sub frame must be a multiple of 4 bytes except for the last sub frame. Padding bytes are not needed for the last MPDU subframe. To make the length of the sub frame in multiple of 4 bytes, padding bytes are appended same as A-MPDU mechanism. Fig. No. 3 shows the structure of an A-MPDU frame. Each MPDU is prefixed with a delimiter. The delimiter contains MPDU length, cyclic redundancy checks (CRC) and unique pattern. The first four bits in the delimiter are reserved and currently not used. 12 bits MPDU Length subfields are used for representing the length of the current MPDU. Cyclic redundancy check calculation includes reserved and length sub fields. The unique pattern is used to find the next delimiter with minimal computation in case of a corrupted delimiter. In this aggregation mechanism selective retransmission is possible due to the presence of individual frame check sequence (FCS) for each MPDU. All the MPDUs in an A-MPDU have same traffic identifier (TID) to effectively work with block acknowledgement (BA) mechanism. The maximum length size of an A-MPDU frame is 64 Kbytes. Capability to receive max size of A-MPDU is different for each station and this capability information is announced at the time of network entry process. Even though an A-MPDU frame can have a maximum size of 64 Kbytes, it can aggregate a maximum of 64 MSDUs. This limitation arises because of the use of block acknowledgment mechanism. To overcome this limitation Two level aggregation is proposed, and it can be used where in multiple A-MSDUs each carrying more than one MSDUs are aggregated to form an AMPDU. In this mechanism the size of A-MSDU frame should not exceed 4 KB.

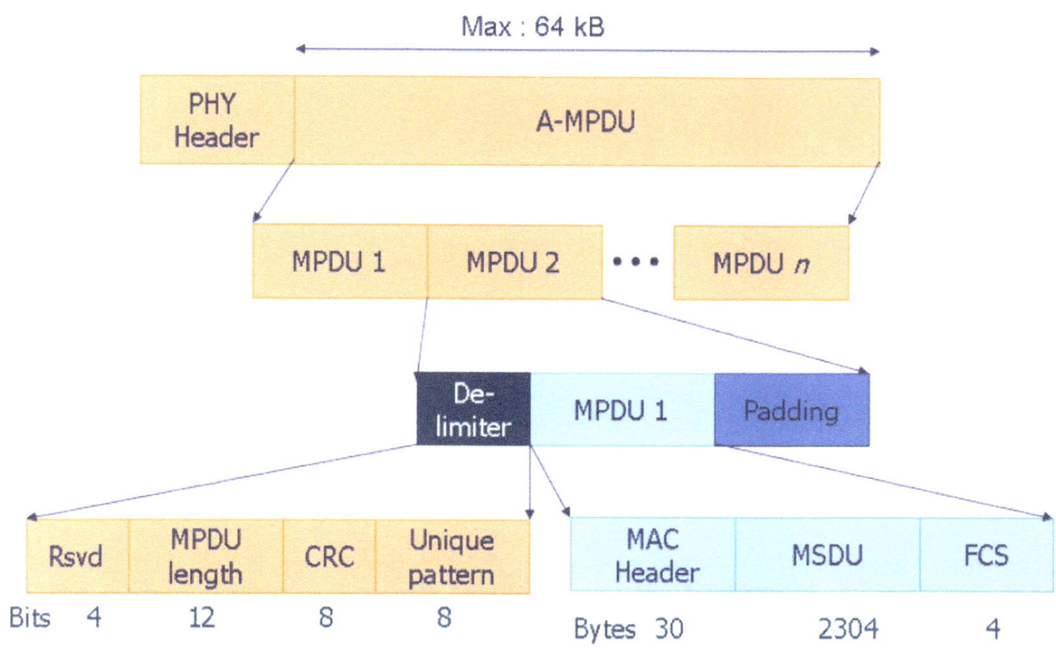

Fig.No. 3 A-MPDU frame format

3.2.4 Two-Level Aggregation

A two-level frame aggregation mechanism comprises a blend of A-MSDU and A-MPDU over two stages. In Fig. 4 it illustrates how this new mechanism can be achieved.

The basic operation is as follows: In the first level, if any MSDUs that are buffered in the A-MSDU provisional storage area justify the A-MSDU constraints explained in the previous related subsection, these data units can be aggregated into a single A-MSDU. If the TIDs are different, all these frames can move to the second level where they will be combined together with any A-MSDUs arises from the first level or other single MSDUs by using A-MPDU aggregation.

However, it must be mentioned that the maximum MPDU length for an A-MPDU data frame is limited to 4095 bytes, and then A-MSDUs or MSDUs with lengths larger than this limitation cannot be transmitted. Conjointly, any frames from an A-MSDU or MSDUs also cannot be included in an A-MPDU. Fig no. 4 illustrates how the two level aggregation mechanism is more efficient in most of the cases than A-MPDU and A-MSDU aggregation operating alone.

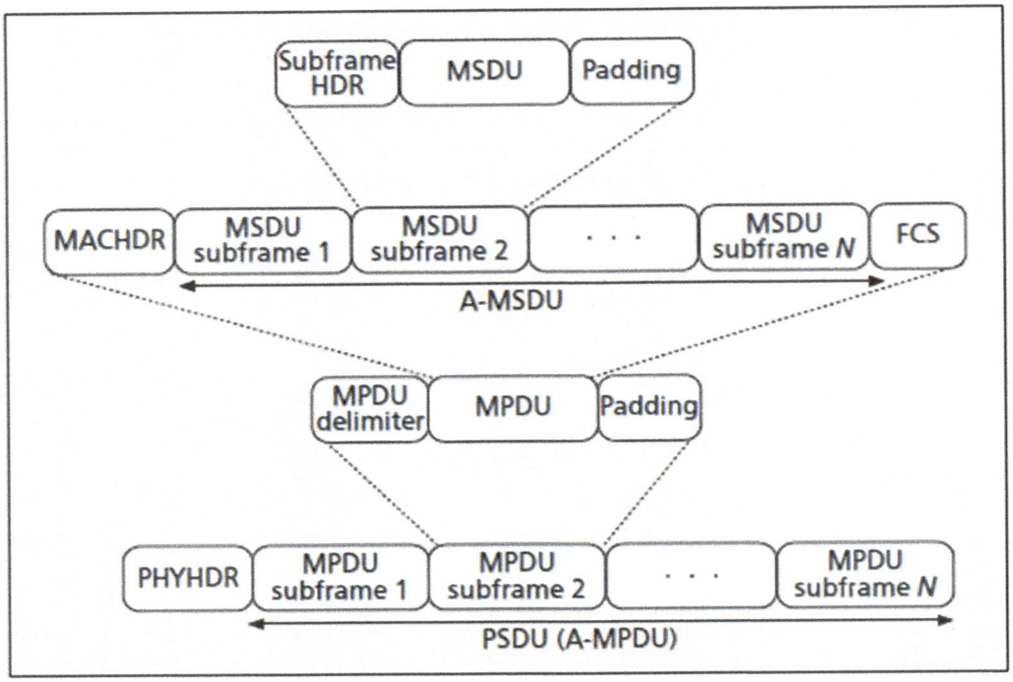

Fig. No. 4: Two Level Aggregation

4. Simulation Setup

In this section, the performance of proposed scheme is evaluated. It compares the performance of the latest draft of the IEEE 802.11n A-MPDU aggregation schemes along with the two-level aggregation. Simulation result shows the different scenario. Proposed scheme is validated using network simulator (i.e. NS2) based on the TCL with the latest 802.11n PHY and MAC enhancements.

Behaviour of a large-scale and complex network system such as Internet at low cost under different configurations of interest and over long period can be predicted by Network simulation software. Many network simulators, such as NS2, Openet, Qualnet, matlab etc., are widely available to simulate network. NS2 is a discrete event simulator written in C++, with an OTcl interpreter as the user interface that allows the input model files (Tcl scripts) to be executed..

The simulator supports a class hierarchy in C++, and a very similar class hierarchy in OTcl. The root of this class hierarchy is the Tcl Object in OTcl. NS2 provides substantial support for simulation of TCP, routing algorithms, queuing algorithms, and multicast protocols over wired and wireless (local and satellite) networks, etc. It is freely distributed, and all source code is available.

NS simulator covers a very large number of applications, protocols, network types, network elements and traffic models. These are called "simulated objects". The simulator is based on two languages: an object oriented simulator written in C++ and an Otcl interpreter used to execute user's command scripts. In these scripts, the user can define a particular network topology, the specific protocols and applications that he wishes to simulate and the form of output he wishes to obtain from the simulator.

NS is a discrete event based simulator. The tcl script defines when an event should occur. For example: $ns at 1.0 "$ftp start". When $ns is a Simulator instance (an event scheduler) and $ftp is a FTP application that was set above lower layers (TCP, physical network – topology, delays, etc.) earlier defined by the user in the script.

Trace file result gives event registered at the network. Event is one of: receive (at output of a link), enqueue, dequeue and drop. Time is the time of the event. From node is the input node of the link at which the event occurs and To node is the output node of that link. Pkt. type (tcp, cbr, ...) and size describe the packet. Fid is the link number that the user can set. Src and Dst addr is the source and destination nodes of the packet. Next is the network layer protocol's sequence number (for instance tcp's sequence number) and the last field holds a unique id of the packet in the system. In addition, the user can define a custom trace format to track system parameters that interest him. For example the queues at the input node of links or the window

Simulation has four scenarios. 20 nodes, 30 nodes, 40 nodes and 50 nodes simulations separately with A-MPDU and Two Level Aggregations frame aggregation mechanisms. Each scenario gives analysis of increased number of nodes separately for two frame aggregation mechanism. In 20 nodes scenario, 20 no nodes topology is created with mac value mac/A-MPDU and AODV routing technique. Here the CBR packet size is assumed of 512 Bytes and it is fixed.

This scenario generates trace file namely aodv.tr and nam file namely aodv.nam. Trace file gives the tracing result of topography, and nam file gives the actual animation of 20 nodes topography. Fig no. shows the trace file of scenario 20 nodes with A-MPDU aggregation. Fig no.5 shows the nam file of scenario 20 nodes with A-MPDU aggregation. Similarly different trace files are generated for different scenario. Fig no.6 shows the trace file of scenario 20 nodes with A-MPDU aggregation Awk scripts are used to analysis different parameters (like average throughput, energy, packet data rate, delay) using these tracing results. All analysed results are collected and represented in xgraph tool of Linux.

5. RESULTS

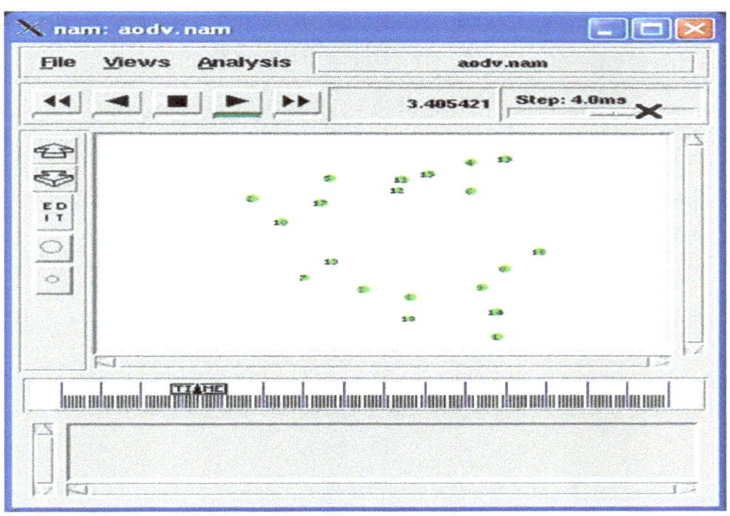

Fig no. 5 Nam file of scenario 20 nodes with A-MPDU aggregation

An A-MPDU aggregation scheme along with the two-level aggregation is compered using different parameters. Following graphs are showing the results for every case in terms of throughput, end to end delay, energy consumption and packet delivery ratio by varying different numbers of nodes on the basis of different parameters.

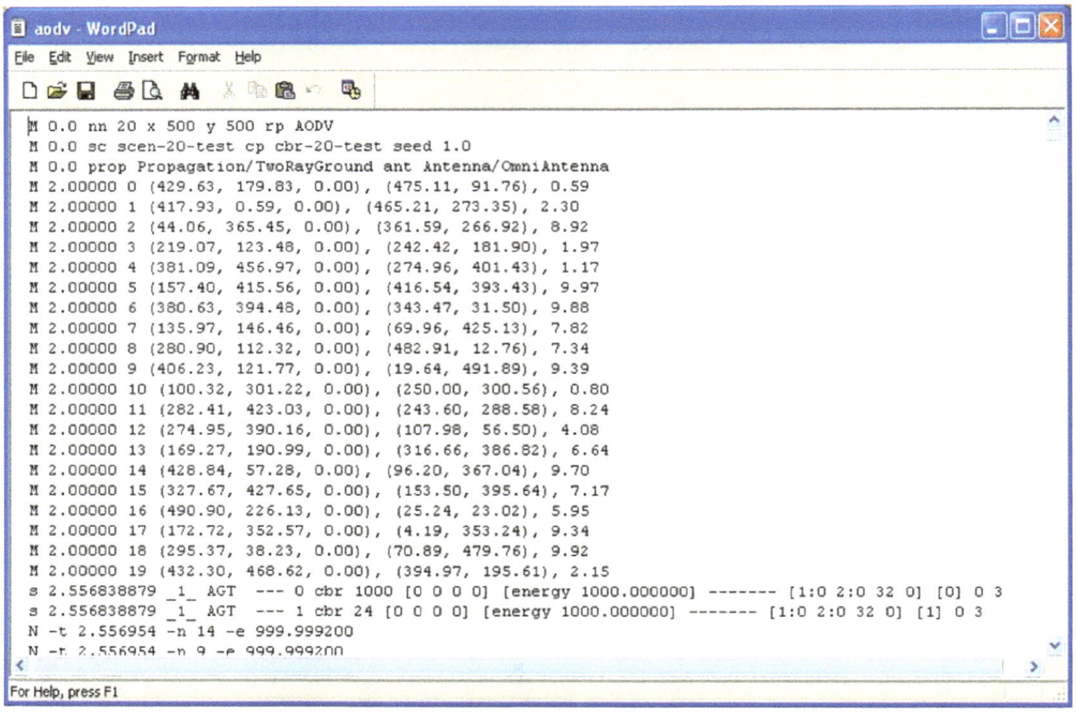


```
M 0.0 nn 20 x 500 y 500 rp AODV
M 0.0 sc scen-20-test cp cbr-20-test seed 1.0
M 0.0 prop Propagation/TwoRayGround ant Antenna/OmniAntenna
M 2.00000 0 (429.63, 179.83, 0.00), (475.11, 91.76), 0.59
M 2.00000 1 (417.93, 0.59, 0.00), (465.21, 273.35), 2.30
M 2.00000 2 (44.06, 365.45, 0.00), (361.59, 266.92), 8.92
M 2.00000 3 (219.07, 123.48, 0.00), (242.42, 181.90), 1.97
M 2.00000 4 (381.09, 456.97, 0.00), (274.96, 401.43), 1.17
M 2.00000 5 (157.40, 415.56, 0.00), (416.54, 393.43), 9.97
M 2.00000 6 (380.63, 394.48, 0.00), (343.47, 31.50), 9.88
M 2.00000 7 (135.97, 146.46, 0.00), (69.96, 425.13), 7.82
M 2.00000 8 (280.90, 112.32, 0.00), (482.91, 12.76), 7.34
M 2.00000 9 (406.23, 121.77, 0.00), (19.64, 491.89), 9.39
M 2.00000 10 (100.32, 301.22, 0.00), (250.00, 300.56), 0.80
M 2.00000 11 (282.41, 423.03, 0.00), (243.60, 288.58), 8.24
M 2.00000 12 (274.95, 390.16, 0.00), (107.98, 56.50), 4.08
M 2.00000 13 (169.27, 190.99, 0.00), (316.66, 386.82), 6.64
M 2.00000 14 (428.84, 57.28, 0.00), (96.20, 367.04), 9.70
M 2.00000 15 (327.67, 427.65, 0.00), (153.50, 395.64), 7.17
M 2.00000 16 (490.90, 226.13, 0.00), (25.24, 23.02), 5.95
M 2.00000 17 (172.72, 352.57, 0.00), (4.19, 353.24), 9.34
M 2.00000 18 (295.37, 38.23, 0.00), (70.89, 479.76), 9.92
M 2.00000 19 (432.30, 468.62, 0.00), (394.97, 195.61), 2.15
s 2.556838879 _1_ AGT  --- 0 cbr 1000 [0 0 0 0] [energy 1000.000000] ------- [1:0 2:0 32 0] [0] 0 3
s 2.556838879 _1_ AGT  --- 1 cbr 24 [0 0 0 0] [energy 1000.000000] ------- [1:0 2:0 32 0] [1] 0 3
N -t 2.556954 -n 14 -e 999.999200
N -t 2.556954 -n 9 -e 999.999200
```

Fig no.6 Trace file of scenario 20 nodes with A-MPDU aggregation.

5.1. Packet Delivery Ratio

Fig. No 6 shows the packet delivery ratio as the function parameter of increasing number of nodes. The ratio of the data packets delivered to the destinations to those generated by the CBR sources is known as packet delivery ratio [32]. The rate of increase of packet delivery ratio for A-MPDU is much lower than for Two level aggregation. Increase in PDR in case of Two level aggregation is due to the large number of frames are received at the sinks.

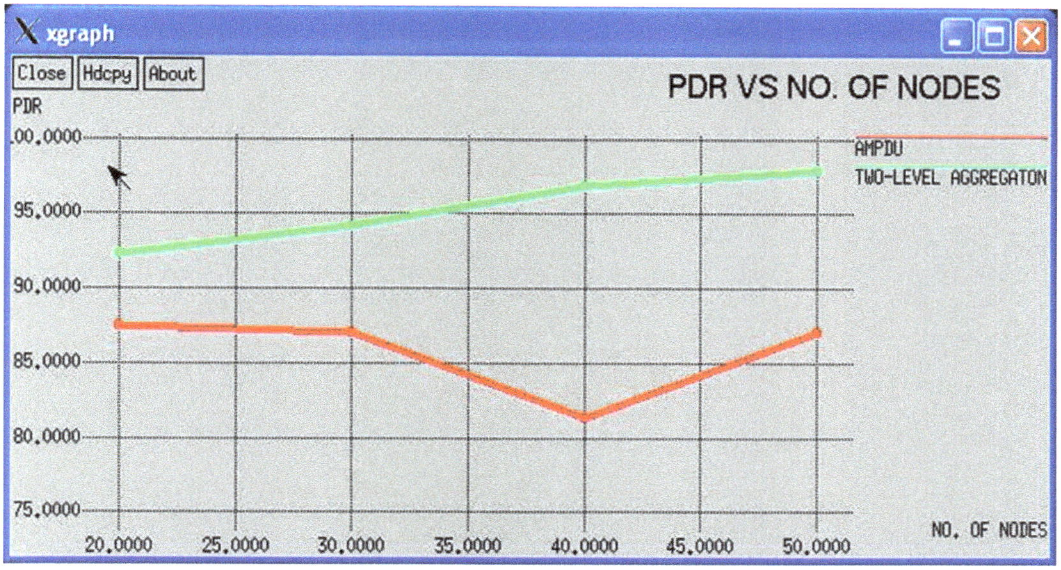

Fig No. 7 Effect of number of nodes on packet delivery ratio

5.2. Average Delay

The delay includes processing and queuing delay in each intermediate node. Average Delay measures the average one way latency observed between transmitting an event and receiving it at each destination. Ideally, Average Delay should have a rather constant value. Figure 8 shows the average delay observed as a function of network size. Increasing rate of the average delay in Two level aggregation is low as compared to that of the A-MPDU due to its second stage aggregation approach in Two level. Average delay of A-MPDU technique is 1 sec at 20 nodes while average delay of Two level aggregation is minimal at 20 nodes compared with A-MPDU

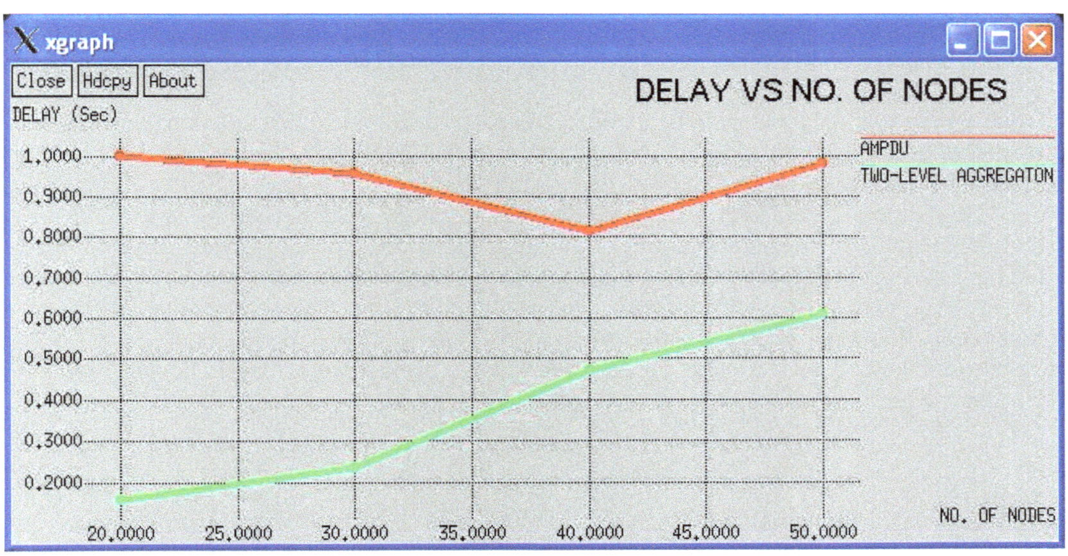

Fig No. 8 Effect of number of nodes on average delay

5.3 Energy Consumption

Fig. No 9 shows the average energy consumed by nodes during the simulation for A-MPDU as well as Two level aggregation scenario. Average energy consumed by A-MPDU technique is 50 Joules, and it's steady for increasing no of nodes. But energy consumed by Two level aggregation technique is 10 joules at 10 nodes and it's slowly increased as numbers of nodes are increased. And at 50 nodes it is only 30 Joules. Simulation result shows that average energy is consumed by Two level aggregation is less than A-MPDU.

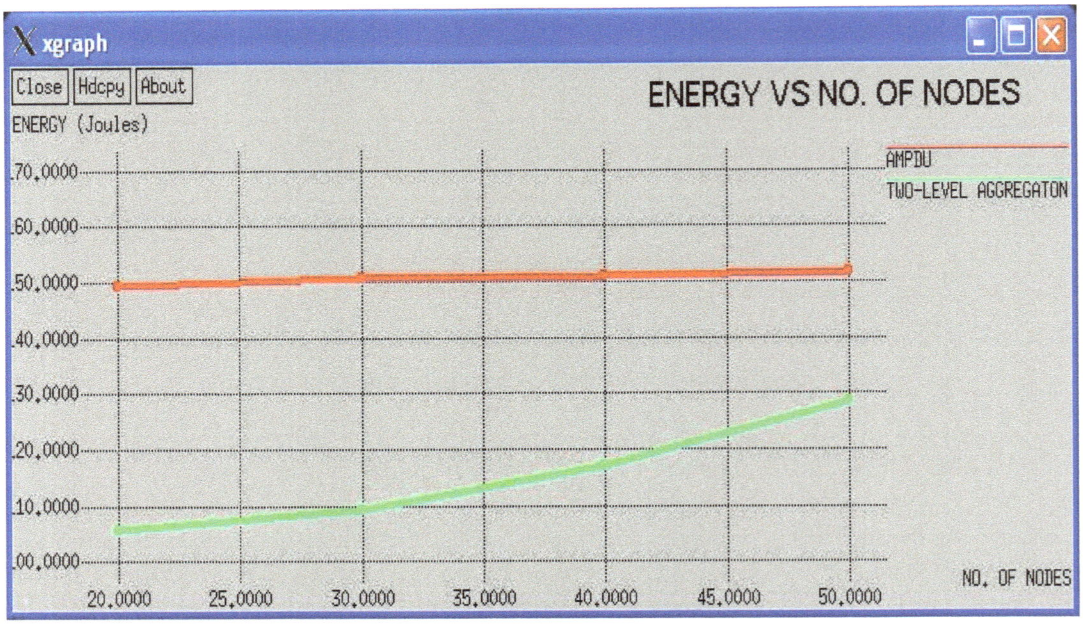

Fig No. 9 Effect of number of nodes on Energy consumed by different scenario

5.4 Throughput

The throughput is the maximum rate at which MAC layer can transmit packets from senders to receivers without packet loss. MAC throughput performance of the schemes increases with an increase of the overload of channel. However, after channel is saturated, the throughput of the A-MPDU scheme is fluctuated. This behaviour is characterized as normal because of channel saturation, meaning the resources are limited to the impending demand. Fig No. 10 shows MAC throughput performance by A-MPDU and Two level aggregation schemes at increasing number of nodes. As number of nodes increases the throughput two-level aggregation is increases exponentially. Throughput achieved by the two-level aggregation is higher than the A-MPDU aggregation scheme. And it is 12Kbps at 50 nodes while throughput of A-MPDU aggregation is fluctuated.

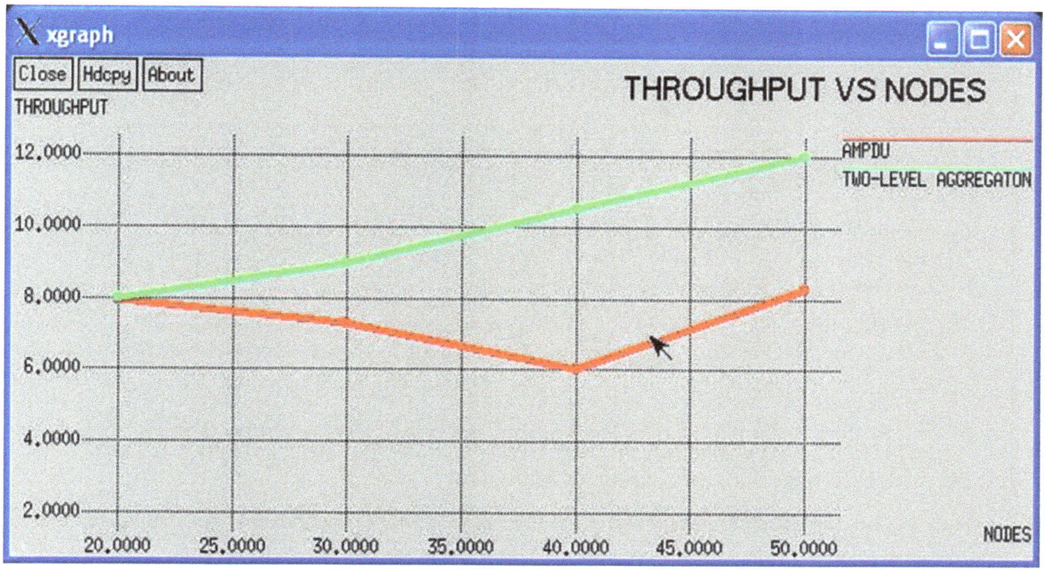

Fig. No. 10 Impact of number of nodes on MAC throughput of different scenario

6. Conclusions

This report gives detailed information about latest IEEE 802.11 draft. Different standards of IEEE 802.11draft are explained with their parameter comparison. Here it gives detailed information of latest IEEE 802.11n WLAN draft. This report explained different techniques which enhance physical as well as MAC layer of IEEE 802.11N amendment. For physical enhancement, MIMO, OFDM, Channel bonding techniques are used. This report highlighted different problems in PHY enhancement, and it resolved by MAC enhancement. In MAC enhancement several mechanisms are explained. From which frame aggregation scheme is proposed in the latest IEEE 802.11n draft. Several frame aggregation mechanism are explained, namely A-MSDU, A-MPDU, and the combined two-level aggregation.

Intensive and detailed study of aggregation techniques and comparison of the performance of the two mechanisms A-MPDU and two level aggregation by network simulator (ns2) was done. The simulation results of comparison have been presented in the form of graphs. Simulation results of this project demonstrated that due to second level aggregation mechanisms large numbers of frames are aggregated in two level aggregation. Two level aggregation can more effectively deliver data from multiple sources PDR rate is much higher with much lower delay and less energy consumption as compared with A-MPDU approach.

As more number of frames are delivered effectively in two level aggregation, rate of increasing data throughput is much higher as compared to A-MPDU aggregation mechanism.

Future work includes comparison between these scenario with additional parameters such as packet size, fragment size and random node mobility. It also includes investigating of the optimal frame size to maximize the throughput of current 802.11n draft; optimal frame size will depend on BER (bit error rate).

REFERENCES

[1] IEEE Std 802.11 WG, "Part 11: Wireless LAN Medium Access Control (MAC) and Physical Layer (PHY) Specifications, Aug. 1999

[2] IEEE 802.11n, Draft 4.0 -: Enhancements for Higher Throughput, IEEE, March 2008

[3] Youngsoo Kim+* Sunghyun Choi+ Kyunghun Jang* Hyosun Hwang* Throughput Enhancement of IEEE 802.11 WLAN via Frame Aggregation supported by Samsung and Information Technology Research Center (ITRC).

[4] Dionysios Skordoulis and Qiang Ni Hsiao-Hwa Chenieee 802.11n mac frame aggregation Mechanisms for next-generation High-throughput wlans 1536-1284/08/$25.00 © 2008 IEEE IEEE Wireless Communications • February 2008

[5] Gautam Bhanage, Rajesh Mahindra, Ivan Seskar, Dipankar Raychaudhuri Implication Of MAC Frame Aggregation On Empirical Wireless Experimentation 978-1-4244-4148-8/09/$25.00 ©2009

[6] Oliver Hoffmann, Member, IEEE, Ruediger Kays, Member, IEEE Efficiency of Frame Aggregation in Wireless Multimedia Networks Based on IEEE 802.11n 2010 IEEE 14th International Symposium on Consumer Electronics

[7] Anwar Saif, Mohamad Othman, Shamla Subramanian, Frame aggregation in wireless Networks: Techniques and Issues. IETE Technical review, vol28 issue 4 , Jul-Aug 2011.

[8] 802.11n: Next-Generation Wireless LAN Technology. April 2006

[9] Boris Ginzburg Alex Kesselman, Intel Corporation Performance Analysis of A-MPDU and A-MSDU Aggregation in IEEE 802.11n

[10] M.Greis. Tutorial for the Network Simulator NS2. [Online]. Available: http://www.isi.edu/nsnam/ns/tutorial/[Accessed on Nov, 2010]

[11] Alexey Sidelnikovx, Jeonggyun Yuy and Sunghyun Choiz Samsung Electronics, Co., LTD, "Fragmentation/Aggregation Scheme for Throughput Enhancement of IEEE 802.11n WLAN"

[12] T.M Nazmul Huda, "Throughput Enhancement of IEEE 802.11 WLAN for Multimedia Communications" 978-1-4244-4570-7/09/$25.00 ©2009 IEEE

[13] Haifeng Zheng, Guotai Chen, and Lun Yu, "Video Transmission over IEEE 802.11n WLAN with Adaptive Aggregation Scheme". IEEE.

[14] The network Simulator Wiki.[Online] Available: http://nsnam.isi.edu/nsnam/index.php/ [Accessed on Nov,2010]

[15] The Network Simulator – ns-2. [Online]. Available: http://www.isi.edu/nsnam/ns/ [Accessed on Nov, 2010].

[16] M.Greis. Tutorial for the Network Simulator NS2. [Online]. Available: http://www.isi.edu/nsnam/ns/tutorial/[Accessed on Nov, 2010]

[17] J. Chung and M. Claypool. Ns by example. [Online]. Available: http://nile.wpi.edu/NS/ [Accessed on Dec, 2010]

18

Analytic Study of Synchronization Errors in OFDM Systems Applied on WLAN Transceiver

Mourad MELLITI[1], Salem HASNAOUI[1] and Ridha BOUALLEGUE[2]

[1]Department of Computer and Communication Technologies
National School of Engineering of Tunis (ENIT)
mourad.melliti@enit.rnu.tn, salem.hasnaoui@enit.rnu.tn
[2]6'Tel Research Unit, Higher School of Communications of Tunis, Sup'Com, Tunisia
ridha.bouallegue@supcom.rnu.tn

Abstract

In this paper, a new energetic approach was developed to study the impact of synchronization errors on OFDM transmission systems. Indeed, this approach can be applied in wireless communication systems such as HiperLAN and WLAN Transceivers. Firstly, the studies were conducted on the impact of timing errors on system performance and we have shown that stress recovery sampling frequency were particularly hard because of the use of a minimum number of pilot symbols and a maximum number of states of the constellations. The mathematical approach we followed in this paper which is an original contribution in the RF field can jointly estimate the channel response and the frequency offset. We realized, along with mathematical analysis, simulations to evaluate the performance of our approach. The results can evaluate the performance of the system against synchronization effects.

Keywords

OFDM, WLAN Transceivers, Frequency Offset, Synchronization effects.

1. Introduction

Several previous studies have investigated and presented general specifications of OFDM systems and identify RF synchronization objectives associated with them. For this reason, it appears that a detailed mathematical analysis of synchronization errors on transmission systems is unavoidable. Indeed, this paper outlines specific problems related to time and frequency offsets at the reception of the OFDM signal which is superimposed with phase noise from a local oscillator located in the receiver. Two phase error effects must be distinguished: interference between carriers and the common phase error. A mathematical analysis will be introduced in this paper to show the quantitative importance of these effects. Among the goals of this work is to demonstrate that the phase noise was not an obstacle in the implementation of a complete digital transmission system and it is essential to carefully evaluate the effects of synchronization errors caused by noise due to channel estimation errors, phase noise, frequency offset, the frame errors and the time lag in the design of receivers.

2. Energetic Analysis of Errors on the Performance Degradation of OFDM Systems

In the following paragraph we will analyze and quantify the various errors that may occur during a transmission OFDM applied to our architecture and caused by a non-ideal synchronization in the placement of the FFT window, the problems of synchronization clock, and the problems of channel estimation. For these studies we illustrate below some concepts to be introduced and some formulas that are necessary to be mentioned and that will be useful throughout the remainder of this paper. Consider the following report introduces the ideal relationship between the ideal SNR and the real SNR and recall that the actual signal to noise ratio means the quality of a transmission of information compared to parasites [1] [2] [3]:

$$\Delta\gamma = \frac{\gamma_{ideal}}{\gamma_{real}} \tag{1}$$

The signal to noise ratio in the case of a perfect synchronization can lead to the following relation while considering that the only disturbances are caused by a Gaussian channel:

$$\gamma_{idéal} = \frac{E_s}{\sigma^2_{AWGN}} \tag{2}$$

While the signal to noise ratio with a non-ideal synchronization is given by the following relation:

$$\gamma_{réel} = \frac{E_s}{\sigma^2_{AWGN} + \sigma^2_{\varepsilon}(E_s;\sigma^2_{AWGN})} \tag{3}$$

While:

$\Delta\gamma$ SNR Loss due to non-ideal synchronization

E_s Averaged received symbol energy

σ^2_{AWGN} Noise power (AWGN)

σ^2_{ε} Is the sum of the different sources of disturbance other than the AWGN as:

$$\sigma^2_{\varepsilon} = \sigma^2_h + \sigma^2_{ph} + \sigma^2_{\Omega} + \sigma^2_t + \sigma^2_{fd} \tag{4}$$

Where:

σ^2_h Due to channel estimation error

σ^2_{ph} Due to phase noise

σ^2_{Ω} Due to frequency offsets

σ^2_t Due to frame and timing errors

σ^2_{fd} Due to fast fading

The report given by equation (3.1) can be reproduced by introducing the concepts of powers of the noise data distributed by equation (3.4) as follows:

$$\Delta\gamma = \frac{\gamma_{idéal}}{\gamma_{real}} = \frac{\sigma^2_{AVGN} + \sigma^2_{\varepsilon}(E_s;\sigma^2_{AWGN})}{\sigma^2_{AVGN}} = 1 + \frac{\sigma^2_{\varepsilon}(E_s;\sigma^2_{AVGN})}{\sigma^2_{AVGN}} \tag{5}$$

Naturally if we raise the ratio of SNR loss in case of bad timing by the maximum loss in SNR as follows:

$$\Delta\gamma \le (\Delta\gamma)_{\max} \tag{6}$$

Then equation (3.5) leads us to increase the following report:

$$\sigma_{\varepsilon}^{2}(E_{s};\sigma_{AVGN}^{2}) \leq \sigma_{AVGN}^{2}.\{(\Delta\gamma)_{\max}-1\} \tag{7}$$

2. Quantification of Synchronization Errors on the Shift Frames

Before addressing the various cases of lag on the arrival of frames at the receiver we will recall the waveform received with cyclic prefix, which can be written as a vector of symbols that is reflecting the simplified expression of the signal at the receiver input and resulting in the following simplified form [4] [6] [7]:

$$y_n = \frac{1}{N}\sum_{k=-K}^{K} X_k H_k e^{\frac{j2\pi nk}{N}} \tag{8}$$

After FFT processing the form of this vector of received symbols becomes:

$$Y_k = \sum_{n=0}^{N-1} y_n e^{\frac{-j2\pi nk}{N}} = X_k H_k \tag{9}$$

Where

$$H_k = \sum_{n=0}^{N-1} h_n e^{\frac{-j2\pi nk}{N}} \tag{10}$$

The time lag of one symbol is expressed by introducing a time error δT_S in the entire duration of the symbol which becomes $(\varepsilon_t + \delta)T_S$ where ε_t is an integer and δ belongs to $[0,1]$.

2.1. Errors for a non-frequency selective channel

Now treat the case where the channel is not frequency selective $\varepsilon_t > 0$ as shown in Figure 1 below:

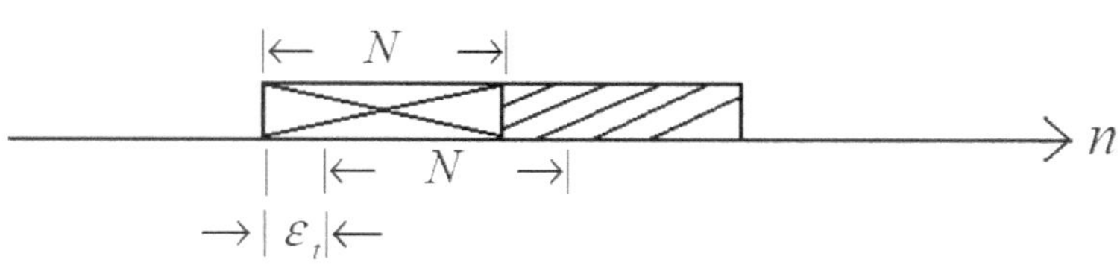

Figure 1. Cas d'un canal non sélectif en fréquence pour $\varepsilon_t > 0$

We then define in the sequel $y_{l,n}$ as the $l^{ième}$block with the $n^{ième}$ sample devoid of the noise caused by the phase shift.

$$Y_{l,k} = \sum_{n=0}^{N-1-\varepsilon_t} y_{l,n+\varepsilon_t} e^{\frac{-j2\pi nk}{N}} + \sum_{n=N-\varepsilon_t}^{N-1} y_{l+1,n-(N-\varepsilon_t)} e^{\frac{-j2\pi nk}{N}} \tag{11}$$

$$Y_{l,k} = \sum_{n=0}^{N-1-\varepsilon_t}\left(\frac{1}{N}\sum_{k=-K}^{K} X_{l,i} e^{\frac{j2\pi i(n+\varepsilon_t)}{N}}\right)e^{-j\frac{2\pi nk}{N}} + \sum_{n=N-\varepsilon_t}^{N-1}\left(\frac{1}{N}\sum_{i=-K}^{K} X_{l+1,i} e^{\frac{j2\pi i(n-N+\varepsilon_t)}{N}}\right)e^{-j\frac{2\pi nk}{N}} \tag{12}$$

$$Y_{l,k} = \overbrace{\frac{N-\varepsilon_t}{N} X_{l,k} e^{\frac{j2\pi k\varepsilon_t}{N}}}^{\text{Part 1}} + \overbrace{\frac{1}{N}\sum_{n=0}^{N-1-\varepsilon_t}\left(\sum_{i=-K,i\neq K}^{K} X_{l,i} e^{\frac{j2\pi i(n+\varepsilon_t)}{N}}\right) e^{-j\frac{2\pi nk}{N}}}^{\text{Part 2}\quad ICI}$$
$$\underbrace{+\frac{1}{N}\sum_{n=N-\varepsilon_t}^{N-1}\left(\sum_{i=-K}^{K} X_{l+1,i} e^{\frac{j2\pi i(n-N+\varepsilon_t)}{N}}\right) e^{-j\frac{2\pi nk}{N}}}_{ISI\quad \text{Part 3}} \tag{13}$$

Where $\frac{N-\varepsilon_t}{N}$ in Part 1 is the coefficient showing the loss of signal and $e^{\frac{j2\pi k\varepsilon_t}{N}}$ is the phase shift term. So the quantization errors on the time lag expressed by calculating the variance of the last two terms Part 2 and Part 3 of the previous relation :

$$\sigma_t^2(E_s;K;N;k) = VAR\left[\begin{array}{l}\frac{1}{N}\sum_{n=0}^{N-1-\varepsilon_t}\left(\sum_{i=-K,i\neq K}^{K} X_{l,i} e^{\frac{j2\pi i(n+\varepsilon_t)}{N}}\right) e^{-j\frac{2\pi nk}{N}} + \\ \frac{1}{N}\sum_{n=N-\varepsilon_t}^{N-1}\left(\sum_{i=-K}^{K} X_{l+1,i} e^{\frac{j2\pi i(n-N+\varepsilon_t)}{N}}\right) e^{-j\frac{2\pi nk}{N}}\end{array}\right] \tag{14}$$

2.2. Errors of a noiseless non-frequency selective channel

In the case of a noiseless non-frequency selective channel we will rely on the assumptions of non-selectivity following:

$$h(t,\tau) = h(t), \quad y_n = x_n h_n, \qquad n = 0\cdots N-1, \quad h_n = h(t)\ \Big|_{t=nT_s}$$

We can write the vector of received symbols as follows:

$$Y_k = \sum_{n=0}^{N-1} x_n h_n e^{\frac{-j2\pi nk}{N}} \tag{15}$$

We replace x_n by its value, and then we can rewrite the vector of received symbols as follows:

$$Y_k = \sum_{n=0}^{N-1} h_n \left[\frac{1}{N}\sum_{m=-K}^{K} X_m e^{\frac{-j2\pi nm}{N}}\right] e^{\frac{-j2\pi nk}{N}} \tag{16}$$

Or:

$$Y_k = \sum_{m=-K}^{K} X_m \left[\frac{1}{N}\sum_{n=0}^{N-1} h_n e^{\frac{-j2\pi n(k-m)}{N}}\right] \tag{17}$$

Otherwise in writing, the vector of received symbols can be separated into two components to make appeared the ICI noise related to the additive white Gaussian noise (AWGN) [1] [3] [9] [10] [11] [13].

$$Y_k = X_k \overline{h_n} + \overbrace{\sum_{m=-K,m\neq k}^{K} X_m \cdot H_{1,k-m}}^{ICI(AWGN)} \tag{18}$$

Where $$\overline{h_n} = \frac{1}{N}\sum_{n=0}^{N-1} h_n \text{ and } H_{1,k-m} = \frac{1}{N}\sum_{n=0}^{N-1} h_n e^{-j\frac{2\pi n(k-m)}{N}}$$

Now energetically, we can simplify the power of the fast fading by calculating the variance of the noise component in the vector of received symbols given by the following relation:

$$\sigma_{fd}^2(E_s;K;k;h) = VAR\left[\sum_{m=-K,m\neq K}^{K} X_m \cdot H_{1,k-m}\right] \qquad (19)$$

2.3. Errors of a noiseless frequency selective channel

In this section we will develop equations for the effect of a noiseless frequency selective channel as in the previous paragraph we quantify the power of synchronization effects related to interference from the temporal selectivity.

Let the following assumptions:

$$y_n = \sum_{l=0}^{L} x_{n-1} h_{n-l}, \quad n = 0 \cdots N-1 \qquad (20)$$

Where $h_{n,l}$ represent the channel response as $h_{n,l} = h(t,\tau)\,\big|_{t=nT_s,\tau=lT_s}$ with the condition: $h(t,\tau) = 0, \quad \tau \geq LT_s, \quad L \leq N_g$

$$Y_k = \sum_{n=0}^{N-1} y_n e^{\frac{-j2\pi nk}{N}} \qquad (21)$$

However, according to the above assumptions and substituting its value is obtained:

$$Y_k = \sum_{n=0}^{N-1}\left(\sum_{l=0}^{L} x_{n-l} h_{n,l}\right) e^{\frac{-j2\pi nk}{N}} \qquad (22)$$

$$Y_k = \sum_{n=0}^{N-1}\left[\sum_{l=0}^{L}\left(\frac{1}{N}\sum_{m=-K}^{K} X_m e^{\frac{-j2\pi(n-l)m}{N}}\right) h_{n,l}\right] e^{\frac{-j2\pi nk}{N}} \qquad (23)$$

$$Y_k = \sum_{m=-K}^{K} X_m \left[\sum_{l=0}^{L}\left(\frac{1}{N}\sum_{n=0}^{N} h_{n,l} e^{\frac{j2\pi n(m-k)}{N}}\right)\right] e^{\frac{-j2\pi lm}{N}} \qquad (24)$$

However, $h_{n,l} = h_l \Rightarrow \frac{1}{N}\sum_{n=0}^{N} h_l e^{\frac{j2\pi n(m-k)}{N}} = 0, \quad m \neq k$

What makes the equation (24) as follows:

$$Y_k = X_k \sum_{l=0}^{L} h_l e^{\frac{-j2\pi lk}{N}} = X_k \cdot H_k \qquad (25)$$

However, $H(m,k) = \sum_{l=0}^{L}\left(\frac{1}{N}\sum_{n=0}^{N} h_{n,l} e^{\frac{j2\pi n(m-k)}{N}}\right) e^{\frac{-j2\pi lm}{N}}$ (26)

This gives a new expression vector symbols as:

$$Y_k = \sum_{m=-K}^{K} X_m \cdot H(m,k) \tag{27}$$

However,

$$H(m,k) = \sum_{n=0}^{N-1}\left(\frac{1}{N}\sum_{l=0}^{L} h_{n,l} e^{\frac{-j2\pi lm}{N}}\right) e^{\frac{j2\pi n(m-k)}{N}} \tag{28}$$

And:

$$H(k,k) = \sum_{l=0}^{L}\left(\frac{1}{N}\sum_{n=0}^{N} h_{n,l}\right) e^{\frac{-j2\pi lk}{N}} = \sum_{l=0}^{L} \overline{h}_l e^{\frac{-j2\pi lk}{N}} \tag{29}$$

Where the final expression vector of received symbols:

$$Y_k = X_k \cdot H(k,k) + \overbrace{\sum_{m=-K, m\neq k}^{K} X_k H(m,k)}^{ICI} \tag{30}$$

Where the energy quantization of the ICI effects caused by selectivity σ_{fd}^2 :

$$\sigma_{fd}^2(E_s;K;k;h) = E\left[\left(\sum_{m=-K, m\neq K}^{K} X_k \cdot H(m,k)\right)^2\right] \tag{31}$$

3. Quantifying of the Local Oscillator Errors

In practice, we use oscillators that are subject to phase noise, a random perturbation of the sine wave phase stationary [7] [12] [14]. Since we will keep the assumption that uses modules of the FFT and IFFT for modulation and demodulation of the OFDM signal and to introduce the notion of error, we will resume the received signal with a phase noise and frequency offset is:

$$y(t) = \sum_{k=0}^{N-1} x_k H_k(t) e^{j(2\pi f_o t + \theta(t))} \tag{32}$$

The carrier can be developed with a frequency offset and phase noise as follows:

$$e^{j(2\pi f_o t + \theta(t))}, \qquad \theta(t) = 2\pi\delta_f t + \varphi(t) \tag{33}$$

Where f_Δ characterizes the frequency offset and $\varphi(t)$ characterizes the phase noise when $|\varphi(t)| << 2\pi$ all assuming that the disturbances caused by the random process are relatively small. Then we can write:

$$e^{j\varphi(t)} \approx 1 + j\varphi(t) \tag{34}$$

Where the carrier will have the following form:

$$e^{j(2\pi f_0 t + \theta(t))} \approx e^{j2\pi(f_0 + \delta_f)t}(1 + j\varphi(t)) \tag{35}$$

Where f_Δ and $\varphi(t)$ characterize the overall effects of the oscillators of the transmitter and receiver δ_f characterized by the use of ppm (part per million) for example: if $f_0 = 2{,}4GHz$ then $\pm\, 3\,ppm \rightarrow \pm 6{,}8\,KHz$

$\varphi_{(t)}$ Is a stationary process broadly characterized by a spectral density function such as:

$$S_\varphi(f) = \frac{F\{R_\varphi(\tau)\}}{C} \tag{36}$$

With $R_\varphi(\tau) = E[\varphi(t)\varphi(t+\tau)]$ Where C: is the carrier power, F {.} is the Fourier transform and E [.] is the expectation which the shape of the spectral density power oscillator types is illustrated in Figure 2. According to the model proposed in [34] we have:

$$S_\varphi(f) = 10^{-c} + \begin{cases} 10 & |f|<f_1 \\ 10^{-(f-f_1)\frac{b}{f_2-f_1}-a} & f_1<f \\ 10^{(f+f_1)\frac{b}{f_2-f_1}-a} & f_1<-f \end{cases} \tag{37}$$

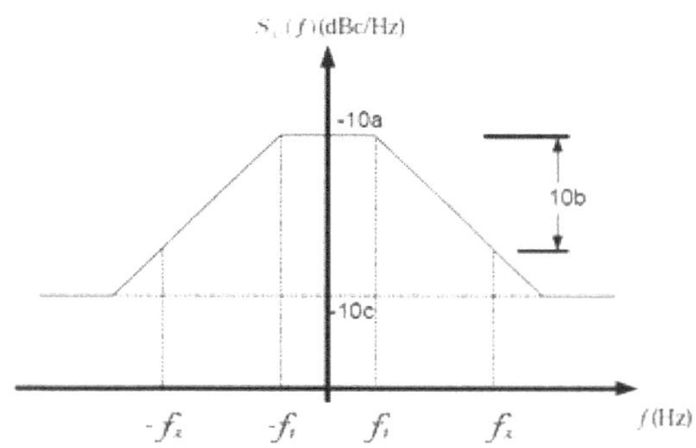

Figure 2. PSD of phase noise in a typical oscillator.

Typical values are: $a = 6.5$, $b = 4$, $c = 10.5$ $f_1 = 1KHz$, $f_2 = 10KHz$ Where C is the background

4. Quantification of Frequency Offset Errors

The OFDM systems are very sensitive to frequency shifts. In Indeed, if such shifts are not compensated, the subcarriers lose their orthogonality and receiver performances are thereby degraded[15] [16] [17] [18]. Quantization errors due to degradation performance due to a frequency offset is developed in following paragraphs.

The components of the vector of received symbols are of the form:

$$Y_k = \frac{1}{N}\sum_{n=0}^{N-1} y_n e^{\frac{j2\pi(k-\varepsilon_\Omega)n}{N}}, \quad when\ \delta_f = -\varepsilon_\Omega \frac{1}{NT_S} \tag{38}$$

By replacing y_n by its expression can then rewrite the vector of received symbols as follows:

$$Y_k = \sum_{n=0}^{N-1}\left[\sum_{l=0}^{L}\left(\underbrace{\frac{1}{N}\sum_{m=-K}^{K} X_m e^{\frac{j2\pi(n-l)m}{N}}}_{x_{n-l}}\right)h_l\, e^{\frac{-j2\pi(k-\varepsilon_\Omega)n}{N}}\right] \quad (39)$$

This can also be written:

$$Y_k = \sum_{m=-K}^{K} X_m\left(\sum_{l=0}^{L} h_l e^{\frac{-j2\pi lm}{N}}\right).\frac{1}{N}\sum_{n=0}^{N-1} e^{\frac{j2\pi(m-k+\varepsilon_\Omega)n}{N}} \quad (40)$$

This gives:

$$Y_k = \sum_{m=-K}^{K} X_m H_m.\frac{1}{N}\sum_{n=0}^{N-1} e^{\frac{j2\pi(m-k+\varepsilon_\Omega)n}{N}} \quad (41)$$

Writing that can be analyzed according to the value of ε_Ω

Where ε_Ω is an integer then we have:

$$\frac{1}{N}\sum_{n=0}^{N-1} e^{\frac{j2\pi(m-k+\varepsilon_\Omega)n}{N}} = \begin{cases} 1 & m+\varepsilon_\Omega = k \\ 0 & si\ non \end{cases} \quad (42)$$

Which gives:

$$Y_k = X_{k-\varepsilon_\Omega} H_{k-\varepsilon_\Omega} \quad (43)$$

Where ε_Ω is not an integer then we have:

$$Y_k = X_k H_k \frac{1}{N}\sum_{n=0}^{N-1} e^{\frac{j2\pi n\varepsilon_\Omega}{N}} + \underbrace{\frac{1}{N}\sum_{m=-K, m\neq k}^{N-1} X_m H_m\left(\sum_{n=0}^{N-1} e^{\frac{j2\pi n(m+\varepsilon_\Omega-k)}{N}}\right)}_{I_k(ICI)} \quad (44)$$

By mathematical simplification we can write:

$$\frac{1}{N}\sum_{n=0}^{N-1} e^{\frac{j2\pi n\varepsilon_\Omega}{N}} = \frac{1}{N}\frac{1-e^{j2\pi\varepsilon_\Omega}}{1-e^{\frac{j2\pi\varepsilon_\Omega}{N}}} \quad (45)$$

This can also be written in another way:

$$\frac{1}{N}\sum_{n=0}^{N-1} e^{\frac{j2\pi n\varepsilon_\Omega}{N}} = \frac{\sin \pi\varepsilon_\Omega}{N\sin\frac{\pi\varepsilon_\Omega}{N}} e^{j\pi\frac{(N-1)\varepsilon_\Omega}{N}} \quad (46)$$

After change of variables we can write:

$$\frac{1}{N}\sum_{n=0}^{N-1} e^{\frac{j2\pi n(m+\varepsilon_\Omega-k)}{N}} = \frac{\sin\pi(m+\varepsilon_\Omega-k)}{N\sin\frac{\pi(m+\varepsilon_\Omega-k)}{N}} e^{j\pi\frac{(N-1)(m+\varepsilon_\Omega-k)}{N}} \tag{47}$$

We can write the increases relationship as follows:

$$\frac{1}{N}\sum_{n=0}^{N-1} e^{\frac{j2\pi n(m+\varepsilon_\Omega-k)}{N}} < \frac{1}{N\sin\frac{\pi(m+\varepsilon_\Omega-k)}{N}} e^{j\pi\frac{(N-1)(m+\varepsilon_\Omega-k)}{N}} \tag{48}$$

Based on the following assumptions and introducing the concept of energy per symbol and always under the assumption of independent white symbols having the same energy, we get[1] [2] [7] [8] [11] [15] [19]:

$$E[X_k]=0 \quad , \quad E[X_k X_l^*]=|X|^2\delta_{l,k}, \quad E[I_k]=0$$

So we get:

$$E\left[|I_k|^2\right] \le |X|^2 \sum_{l=-K-k,l\neq 0} E\left[|H_l|^2\right]\left(\frac{1}{N\sin\frac{\pi(l+\varepsilon_\Omega)}{N}}\right)^2 \tag{49}$$

For an ideal channel, we can write:

$$E\left[|I_k|^2\right] \le |X|^2|H|^2 \sum_{l=-K-k,l\neq 0} E\left[|H_l|^2\right]\left(\frac{1}{N\sin\frac{\pi(l+\varepsilon_\Omega)}{N}}\right)^2 \tag{50}$$

$$E\left[|I_k|^2\right] = \sigma_{\varepsilon_\Omega}^2(E_s, K, k) \tag{51}$$

Numerically, it was determined in [37] that the sum shown in equation (50) is increased by 0.5947 for this case can be rewritten as:

$$E\left[|I_k|^2\right] \le 0.5974|X|^2|H|^2(\sin\pi\varepsilon_\Omega)^2, \quad |\varepsilon| \le 0.5 \tag{52}$$

$$SNR \ge \frac{|X|^2|H|^2(\frac{\sin\pi\varepsilon_\Omega}{\pi\varepsilon_\Omega})^2}{0.5974|X|^2|H|^2(\sin\pi\varepsilon_\Omega)^2 + N_0} \tag{53}$$

$$SNR = \frac{E_s(\frac{\sin\pi\varepsilon_\Omega}{\pi\varepsilon_\Omega})^2}{N_0(1+0.5974\cdot\frac{E_c}{N_0}(\sin\pi\varepsilon_\Omega)^2} \tag{54}$$

$E_s = |X|^2|H|^2$: is the average energy received. Equation (54) gives directly

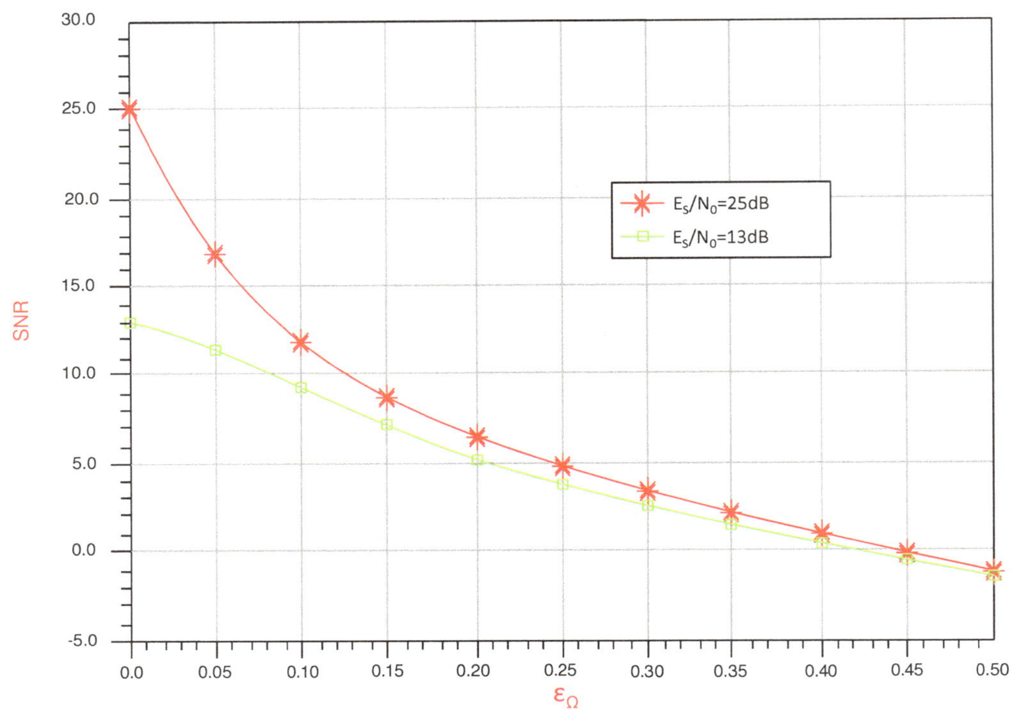

Figure 3. Degradation of signal to noise ratio due to frequency offset, for N = 256, K= 96 for two values of 13dB and ES/N0 be 25 dB.

5. Conclusion

In this paper we presented several approaches for time and frequency synchronization on the OFDM transmission systems which can be applied on wireless communication systems such as HiperLAN WLAN. First, studies that we conducted on the impact of errors synchronization on system performance have shown that stress recovery sampling frequency were particularly harsh because of the use of a number Minimum pilot symbols and a maximum number of states constellations. The mathematical approach that we followed during this chapter provides a joint estimate of the channel response and frequency shift of the clock which is an original contribution. We conducted parallel to mathematical analysis, simulations to evaluate relative performance approaches presented in terms of estimators of the frequency approach, the results obtained allow devalue the performance of the system to synchronization effects.

6. References

[1] U. Tureli, D. Kivanc, and H. Liu, "Experimental and analytical studies on a high resolution OFDM carrier frequency offset estimator, "IEEE Trans. Vehic. Technol, vol. 50, no. 2, pp. 629–643, March 2001.

[2] M. Mourad, H. Salem, and B. Ridha, "Analysis of frequency offsets and phase noise effects on an OFDM 802.11 g transceiver," in Proceedings of the 4th International Workshop on Wearable and Implantable Body Sensor Networks (BSN '07), vol. 13, Aachen, Germany, March 2007.

[3] R. Nogueroles, M. Bossert, A. Donder, andV. Zyablov, "Improved performance of a random OFDM A mobile communication system," in IEEE 48th Vehicular Technology Conference (VTC'98), Ottawa, Canada, May18-21, 1998, vol.3, pp. 2502-2506.

[4] Van der Perre L, Thoen S, Vandenameele P, "Adaptive loading strategy for a high speed OFDM-based WLAN", IEEE Globecom '98, Sydney, Australie, November 1998.

[5] http://eesof.tm.agilent.com

[6] S. K. Bassam, "Frequency Synchronization Errors and Effects of Phase Noise on OFDM Systems," EE359-Wireless Communications Class Projects Stanford University, 2004.

[7] Huang L. C. Tjhung T. T Lu J. "On the optimum number of OFDM subcarriers in fast Rayleigh fading channels". Journal of communication and networks (J. commun. netw.) ISSN 1229-2370. 2001, vol. 3, no2, pp. 142-147.

[8] G. Pujolle, Les Réseaux, Eyrolles, Paris, France, 4th edition, 2000.

[9] G. Bhatia V. Kumar "Adapting MAC 802.11 for performance optimisation of manet using cross layer interaction" International Journal of Wireless & Mobile Networks (IJWMN) Vol.2, No.4, November 2010

[11] T. Oh, Y.B Choi, J Ryoo, K Stokes "Security management in Wireless Sensor Networks for healthcare". International Journal of Mobile Communications 2011 - Vol. 9, No.2 pp. 187 - 207

[12] H. Salem, "Contribution à la modélisation des câbles monotones par éléments finis," thesis report ENIT, January 2000.

[13] S. Lim, C. Yu, C. R. Das "A realistic mobility model for wireless networks of scale-free node connectivity ". International Journal of Mobile Communications. Volume 8 - Issue 3 – 2010.

[14] A. R. S. Bahai and B. R. Saltzberg, Multi-Carrier Digital Communications. Theory and Applications of OFDM, Kluwer Academic Publishers/Plenum, New York, NY, USA, 1999.

[15] Shia-Sheng P, "WLAN IEEE802.11a Transceiver, Algorithm, Architecture and Simulation Results", Report of research, November 2001.

[16] V. S. Abhayawardhana and I. J. Wassell, "Common phase error correction for OFDM in wireless communication," in Proceedings of the 3rd International Symposium on Communication Systems, Networks and Digital Signal Processing (CSNDSP '02), Staffordshire, UK, July 2002.

[17] G.L. Stuber, J.R. Barry, S. W. McLaughlin, Y. Li, M.A.Ingram, and T.G. Pratt, "Broadband MIMO-OFDM wireless communications, "Proceedings of the IEEE, vol. 92, no. 2, pp. 271–294,February 2004.

[18] D. Petrovic, W. Rave, and G. Fettweis, "Phase noise suppression in OFDM using a Kalman filter," in Proceedings of the 6th International Symposium on Wireless Personal Multimedia Communications (WPMC '03), vol. 3, pp. 375–379, Yokosuka, Japan, October 2003.

[19] D. B. Leeson. A simple model of feedback oscillator noise spectrum. Proc. of the IEEE. pp. 329-330, Feb 1966.

19

AN IMPROVED DECENTRALIZED APPROACH FOR TRACKING MULTIPLE MOBILE TARGETS THROUGH ZIGBEE WSNS

Tareq Alhmiedat[1], Amer O. Abu Salem[2], Anas Abu Taleb[3]

[1]Department of Computer Science, Zarqa University, Zarqa, Jordan
t.alhmiedat@zu.edu.jo
[2]Department of Computer Information Systems, Zarqa University, Zarqa, Jordan
abusalem@zu.edu.jo
[3]Department of Computer Science, Isra University, Amman, Jordan
anas.taleb@ipu.edu.jo

ABSTRACT

Target localization and tracking problems in WSNs have received considerable attention recently, driven by the requirement to achieve high localization accuracy, with the minimum cost possible. In WSN based tracking applications, it is critical to know the current location of any sensor node with the minimum energy consumed. This paper focuses on the energy consumption issue in terms of communication between nodes whenever the localization information is transmitted to a sink node. Tracking through WSNs can be categorized into centralized and decentralized systems. Decentralized systems offer low power consumption when deployed to track a small number of mobile targets compared to the centralized tracking systems. However, in several applications, it is essential to position a large number of mobile targets. In such applications, decentralized systems offer high power consumption, since the location of each mobile target is required to be transmitted to a sink node, and this increases the power consumption for the whole WSN. In this paper, we propose a power efficient decentralized approach for tracking a large number of mobile targets while offering reasonable localization accuracy through ZigBee network.

KEYWORDS

Localization, Tracking, Decentralized, Wireless Sensor Networks, ZigBee

1. INTRODUCTION

In recent years, advances in signal processing have led to manufacturing small, low power, inexpensive Wireless Sensor Network (WSN). The signal processing in WSN is different from the traditional wireless networks in two critical aspects: first, the signal processing in WSN is performed in a fully distributed manner, unlike in traditional wireless networks. Second, due to the limited computation capabilities of sensor networks, it is essential to develop an energy and bandwidth efficient signal processing algorithms.
A sensor network is composed of sensor nodes which are small in size, low in cost, and have short communication range. A sensor node usually consists of four sub-systems as follows:

1. Computing subsystem, which is responsible for functions such as execution of the communication protocols and control of sensors

2. A sensing subsystem, that is responsible for sensing the environmental characteristics, such as temperature and humidity

3. A communication subsystem, this consists of a short radio range used to communicate with neighbouring nodes

4. A power supply subsystem, which includes a battery source that provides energy to sensor node.

WSN technology is exciting with unlimited potentials in various application areas including: environmental, medical, military, transportation, homeland defence, crisis management, entertainment, and smart spaces [1, 2, 3, 4, 5]. Researchers have focused on diverse aspects of WSN, such as hardware design, routing, data aggregation and localization [6, 7]. Recently, WSN-based localization and tracking issues have received much attention, driven by the need to accomplish high localization accuracy with the minimum cost, this is because:

1. In many applications, the location itself is the information of interest,

2. Several routing protocols are based on the sensor nodes' locations,

3. Transferring sensors' measurements without incurring the sensors' locations is an unproductive task.

The authors of this paper focused on diverse aspects of tracking mobile targets through distributed sensor networks, such as the localization method [8], tracking multiple mobile targets through ZigBee networks [9], and the data aggregation and prediction methods [10, 11]. In this paper, we focus on the communication between nodes when tracking a large number of mobile targets through ZigBee WSNs, which further intends to reduce the power consumption for WSNs.

WSNs based Localization systems have been researched and addressed extensively in several works [12, 13, 14]. In real time tracking applications, it is essential to continuously transmit the mobile targets' locations to a sink node, in order to display its current position online. Tracking a large group of mobile targets through hundreds of sensor nodes requires the transmission of the location for each mobile target from reference nodes to the sink node, which normally results in a series of hops through the network. Each of these hops increases the consumption of the limited energy, and therefore participates failures within the network as the energy of the reference nodes becomes increasingly impaired.

In this paper, we categorize the tracking systems into centralized and decentralized. In centralized tracking systems, localization information might be transmitted to a sink node to obtain the localization information, while in decentralized systems; localization information is obtained from each mobile target itself, and then transmitted to a sink node. Figure 1 depicts the main idea for both systems (centralized and decentralized). As presented, assume a mobile target is in the range of 4 reference nodes. In centralized systems, a total number of 4 reference messages would be transmitted to the sink node, whereas in the decentralized systems, only a single message that includes the current location of the mobile target, is transmitted to the sink node, as processing the mobile target's location would take place at the reference or the mobile target node itself.

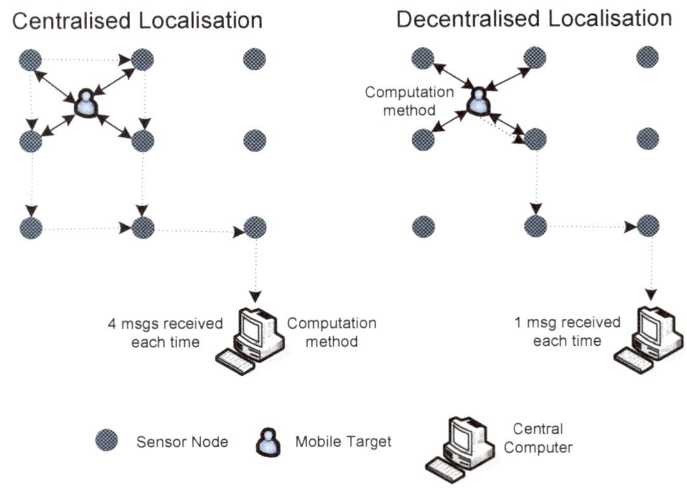

Figure 1. Centralized vs. decentralized localization methods

Tracking a few number of mobile targets through a decentralized approach offers low power consumption [10]. However, when it comes to tracking a large number of mobile targets through distributed sensor nodes, then high power consumption is expected. For that reason, this paper presents a new efficient tracking system, which basically reduces the power consumption required to track multiple mobile targets through ZigBee WSNs. In this paper, we will refer to the following terms:

- *Reference node (r)*: the node with fixed known 2-d coordinates,
- *Mobile target (t)*: the node with unknown location, which requires estimating the location of itself,
- *Group leader (g)*: the *mobile target node* with the minimum number of hops to a sink node is selected to be a group leader in order to aggregate the localization information obtained from adjacent mobile targets, and transmit to a sink node.

In this paper, our contribution lies on the following aspects:

1. Study the existing tracking systems, and categorize them into: centralized and decentralized systems,
2. Design, develop, and implement a power efficient decentralized tracking system using ZigBee WSNs, which intends to minimize the power consumption when tracking a large number of mobile targets,
3. Evaluate the proposed system in this paper, and compare it with the existing systems proposed recently.

2. RELATED WORK

In this section, we summarize the existing centralized and decentralized tracking systems. First, we start with centralized tracking systems. According to [15], centralized based systems have two advantages over the decentralized ones: first, reference nodes are not required to have processing capabilities. Second, the irregularity and mobility of the mobile target requires frequent data communications among reference nodes to have good localization accuracy, which destroys the main advantage of decentralized localization systems.

The works presented in [16, 17, 18] include a RSS based localization system through centralized communication among sensor nodes. In such systems, localization information is transmitted to a sink node in order to process and compute the localization coordinates for each mobile target at the sink node. The work presented in [19] investigated the performance of centralized cooperative positioning algorithm, in which all information has to be collected at a central entity for positioning target nodes. The system proposed in [20] includes a novel computing architecture for WSNs to overcome the Kalman Filter drawbacks, which divides the positioning system into three components: measurement, pre-processing, and data processing.

Second, decentralized or distributed localization techniques; which include that each sensor node is responsible for determining its location with only limited communication with nearby nodes. Decentralized systems have been widely used due to the low energy required compared to the centralized systems. However, this kind of systems requires attaching a localization and computation mechanism to each reference node. In [21], a hybrid technique called RObust Position Estimation (ROPE) was proposed, which allows sensors to locate their locations with no need to a centralized computation facility. ROPE provides a location verification mechanism that verifies the location claims of the sensors before data collection. Thus, ROPE allows the sensors to estimate their own location without the assistance of a central authority.

The system proposed in [22] involves a MDS-MAP technique for calculating the positions of nodes with only basic information that is likely to be already available. MDS-MAP technique involves starting with the given network connectivity information, and an all-pairs shortest-paths algorithm is used to assess the distance between each possible pair of nodes. A decentralized tracking system based on a camera sensor device is proposed in [23] which intended to reduce the bandwidth requirements. The images captured are required to be processed at each camera sensor node with the objective of the extracting location of mobile targets. The proposed approaches in [24, 25, 26] are decentralized based localization systems, since all the communication and processing are undertaken in the sensor node itself.

3. AN IMPROVED DECENTRALIZED TRACKING APPROACH

In WSN systems, long lifetime requirement of sensor nodes has led us to find out new horizons for reducing the power consumption upon nodes. Consuming less energy in tracking applications is a primary objective in designing WSN systems, as each sensor node is usually supported by batteries which could be difficult to replace. The proposed decentralized approach includes grouping the mobile targets in a given area, and selecting a group leader for each group of mobile targets. The group leader is responsible for:

1. Collecting localization information from other mobile targets in its range,
2. Aggregating and transmitting the received localization information to a sink node.

The proposed system in this paper is based on decentralized localization computation. Thus, localization information is processed at the mobile target device, in which each mobile target has the capability to compute its current location using the system proposed in [6]. Only the final coordinates are transmitted to a sink node. This reduces the amount of messages transmitted over the network, and hence shrinks the power-consumption for each reference node. In this section, we discuss the proposed decentralized tracking system, which consists of 3 main phases; initialization, localization, and grouping. Figure 2 depicts the idea of the improved decentralized approach.

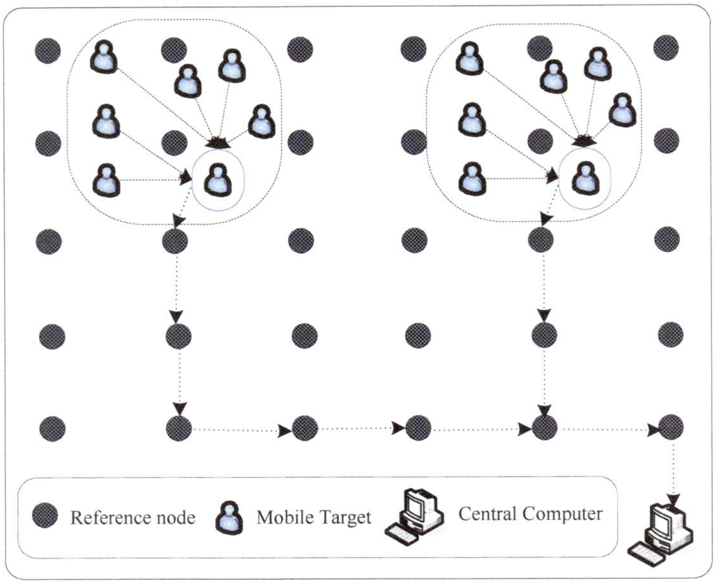

Figure 2. The main idea of the improved decentralized approach

3.1. Initialization phase

This phase includes starting the network, and assigning a network address to reference nodes in the tracking area of interest. Then, mobile targets enter the tracking area and obtain network addresses. Afterwards, each mobile target checks the reference nodes in its range, in order to position itself when required.

3.2. Localization phase

In this phase, each mobile target localizes itself based on the RSS values obtained from adjacent reference nodes. The localization approach proposed in [6] is deployed as a localization method to estimate the mobile targets' locations. This process is repeated every f seconds to estimate the mobile targets' locations simultaneously.

3.3. Grouping phase

This phase involves selecting a group leader for the mobile targets set in the same transmission range. The main reasons behind choosing a group leader is to reduce the amount of messages transmitted to the sink node over the WSN, and hence conserving the amount of energy consumed during the tracking process.

To illustrate this phase, assume $R=\{r_1, r_2, r_3, ..., r_c\}$ represents the reference nodes in the tracking area of interest, with total number *c*, and $T=\{t_1, t_2, t_3, ..., t_e\}$ represents the mobile targets with total number *e*. Assume that $S=\{t_1, t_2, ..., t_z\}$ is a set of mobile targets in a given area set within the same transmission range, with a total number *z*, and each mobile target t is covered by at least three reference nodes to triangulate its current position. Each mobile target checks the number of hops between itself and the sink node. The mobile target with the least number of hops to the sink node is elected to be a group leader. The group leader should have the following characteristics:

1. It has the minimum number of hops to the sink node,
2. It has a reasonable amount of energy, over a certain threshold.

The group leader receives the localization information from other mobile targets in its range, then aggregates, and transmits this information to the sink node. The flowchart for the algorithms running on a mobile target, and the group leader nodes are presented in Figures 3 and 4, respectively.

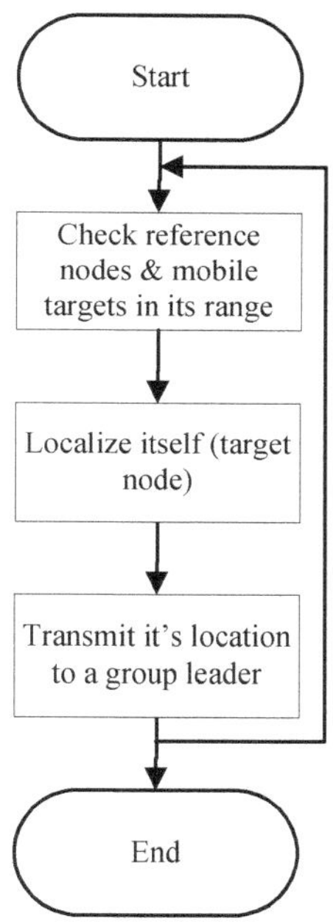

Figure 3. The algorithm implemented on each mobile target node

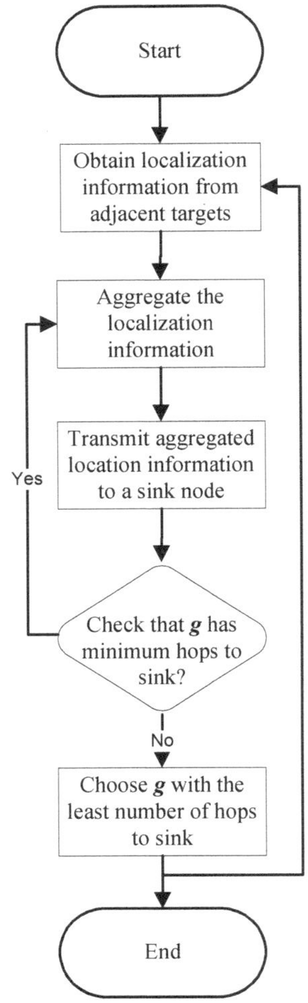

Figure 4. The algorithm implemented on the group leader node

4. SIMULATION EXPERIMENTS

The proposed system is evaluated through implementation using NS2 simulator environment. In order to evaluate the proposed approach in this paper, a total number of three different approaches have been implemented and tested. First, a centralized localization approach, which includes transmitting the localization information from each reference node covering the mobile target to a sink node. Second, a decentralized tracking system, this includes that each mobile target estimates its current position and transmits it to a sink node. And third, an improved decentralized approach, which involves selecting a group leader which placed close to the sink node in order to transmit the localization information through the minimum number of hops.

The system proposed in this work was implemented using ZigBee network standard. ZigBee is a low power consumption, low data rate, and low cost wireless communication standard designed to be used in home automation and remote control applications. ZigBee network standard performs three roles: coordinator, router, and end-device. Only one ZigBee coordinator is required for each WSN, it starts a network formation. ZigBee router is an optional network component; it participates in multi-hop routing of messages. While ZigBee end-device is utilized for low power operation, and is not allowed to participate neither in association nor routing. In this project, we used a ZigBee coordinator as a sink node and a number of ZigBee routers as reference and mobile target nodes.

4.1. Reference and mobile target platform

There are two different types of devices in our test-bed: reference nodes and mobile targets. Reference nodes include stationary sensor nodes with known positions distributed over the tracking area of interest, whereas the mobile targets have the ability to localize themselves. Both, reference and mobile targets were considered as router nodes. Table 1 presents simulation parameters used in our simulation experimental test-bed.

Table 1. Simulation parameters

Parameter name	**Parameter value**
Packet size	127 bytes
InitEng	27.00 *mA.h*
rxPower	49.00 *mA.h*
txPower	44.00 *mA.h*
Simulation time	360 seconds
Number of nodes	56 nodes
Number of mobile targets	10
Average hops	5
Radio model	TwoRayGround
Antenna type	OmniAntenna
Grid size	75 x 65 m^2
Routing protocol	AODV
Mac protocol	MAC/802.15.4

4.2. Reference and mobile target platform

Our experiment test-bed consists of 56 reference nodes distributed over the tracking area of interest, and a sink node placed on the right corner as depicted in Figure 5.

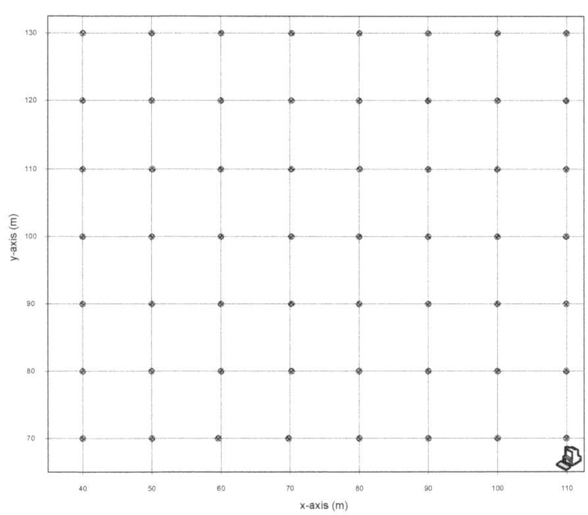

Figure 5. The experimental test-bed

5. SYSTEM EVALUATION

In this section, three scenarios are examined (centralized, decentralized, and improved decentralized), based on evaluating two performance metrics: the total number of messages transmitted over the network, and the power-consumption.

5.1. Communication cost

One of the critical issues in developing a tracking system for sensor nodes is the issue of battery lifetime. Sensor devices are designed to work for periods of several months to a few years. The system proposed in this paper aims to minimize the total number of messages transmitted over the network in order to reduce the amount of energy consumed. In this section, we evaluate the total number of messages transmitted to the sink node for the three different scenarios discussed above (centralized, decentralized, and improved decentralized).

In centralized localization approach, every reference node senses the mobile target has to transmit its readings to a sink node. The total number of messages transmitted over the network is represented in Equation 1.

$$n_1 = r \times m \times \frac{l}{f} \tag{1}$$

where r, m, l and f refer to the number of reference nodes, the number of mobile targets located in one transmission range, total experiment time, and frequency rate (how often the mobiles' locations have to reach the sink node) respectively. For instance, assume the total experiment time is $l = 60$ seconds, the number of reference nodes $r = 3$, and the number of mobile targets is $m = 10$, then the total number of messages transmitted to the sink node is $n_1 = 900$ (where $f = 2$ seconds).

On the other hand, in the basic decentralized approach, two types of messages are transmitted over the network. First, the messages transmitted between reference nodes and mobile targets represented in Equation 1. Second, the messages transmitted between mobile targets and a sink node, as represented in Equation 2.

$$n_2 = m \times \frac{l}{f} \tag{2}$$

As in the previous example, consider the total experiment time is $l = 60$ seconds, the number of reference nodes $r = 3$, and the total number of mobile targets is $m = 10$. Then the average number of local messages will be $n_2 = 900$. These messages are not required to reach a sink node, since they are required to be exchanged between reference nodes and mobile targets. Whereas the total number of messages sent to the sink node is 300 messages. Consequently, there is a significant improvement in the decentralized over centralized approaches.

The improved decentralized approach proposed in this paper, deals with three types of messages transmitted over the network, are as follows:

1. Local messages: indicates messages exchanged between each mobile target and reference nodes, which represented in Equation 1.
2. Group messages: represents messages transmitted between the group leader and other mobile targets in its range. This kind of messages does not affect the energy of reference nodes (wireless sensor nodes), as communications are processed between the mobile targets themselves, which represented in Equation 3.
3. Global messages: represents total number of messages transmitted between the group leader and the sink node, as represented in Equation 4.

$$n_3 = (m-1) \times \frac{l}{f} \quad (3)$$

$$n_4 = \frac{m}{5} \times \frac{l}{f} \quad (4)$$

In equation 4, the total number of messages is divided by 5, as each packet could contain the locations and network addresses for a maximum number of 5 mobile targets. For the same example mentioned above, assume the total experiment time is $l = 60$ seconds, the number of reference nodes $r = 3$, and the number of mobile targets is $m = 10$. Then the total number of local messages is $n_1 = 600$, the total number of group messages is $n_3 = 180$, and the total number of global messages is $n_4 = 40$.

As discussed above, the improved decentralized approach does not require a large number of messages to be exchanged between mobile targets and reference nodes, even when a large number of mobile targets are required to be positioned. Calculations are taken place at the mobile target side, with no need for centralized computations. Figure 6 shows a comparison of the messages required to be transmitted among the three scenarios. The centralized approach requires transmitting a high number of messages between reference and sink nodes, whereas the implemented decentralized approach requires exchanging less messages when the total number of mobile targets is small. At the meanwhile, the improved decentralized approach achieves the least number of messages required to be transmitted over the network mainly when the density of mobile targets is high.

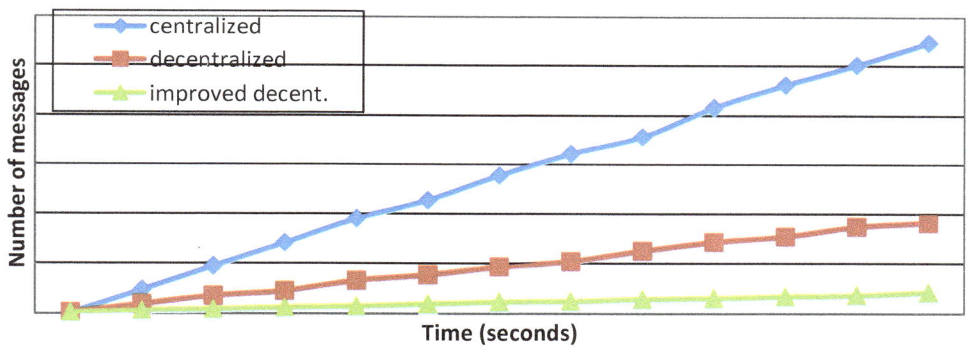

Figure 6. Total number of messages for the three systems

In this work, we aimed to reduce the amount of messages exchanged between wireless sensor nodes in the tracking area of interest. Figure 7 presents the total number of messages exchanged between mobile targets in the three scenarios. As presented, the improved decentralized approach achieves a few more number of messages than centralized and decentralized systems.

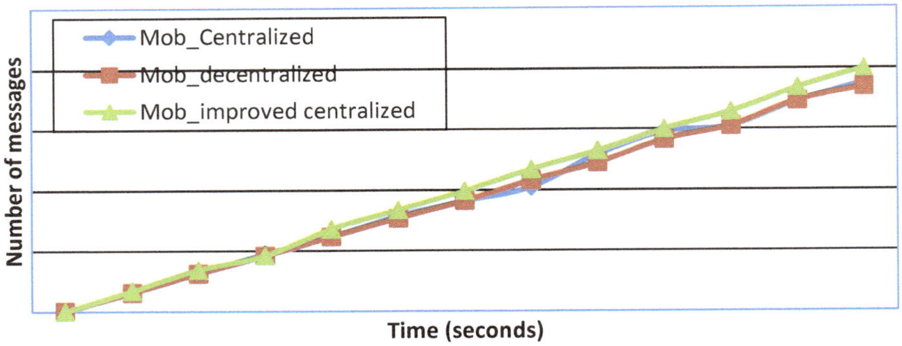

Figure 7. Total number of messages exchanged between mobile targets for the three systems

5.2. Power-consumption

The previous section evaluates and compares the total number of transmitted messages for the three scenarios. Each transmitted message requires a particular amount of power. In this section, the power-consumption for each approach is evaluated. The power consumption for the centralized E_{cn}, decentralized E_{dc} and improved decentralized E_{imp} represented in the Equations 8, 9, and 10, respectively.

$$P(n) = (n \times P_{tx}) + (n \times P_{rx}) \tag{7}$$

$$E_{cn} = P(n_1) \times h \tag{8}$$

$$E_{dc} = \left(P(n_1) + P(n_2)\right) \times h \tag{9}$$

$$E_{imp} = (P(n_1) + P(n_4)) \times h \tag{10}$$

where $P(n)$ is the energy required to transmit a number of packets n, P_{tx} is the amount of energy required to transmit a single packet, P_{rx} is the amount of energy required to receive a single packet, and h is the number of hops. Figure 8 compares the battery life of the three approaches (centralized, decentralized, and improved decentralized). As shown, the improved decentralized method achieves longer battery life than both the centralized and decentralized approaches, as it requires less number of messages to be exchanged between mobile targets and a sink node.

In Figure 9, the power consumption for the group of mobile targets (m = 10) are estimated in three systems (centralized, decentralized and improved decentralized). As shown, the improved decentralized system achieves the highest power consumption. However, there is a slight difference in power consumption between the improved decentralized, centralized and decentralized systems. The mobile target devices are easy to maintain in case if any target device's battery has run out.

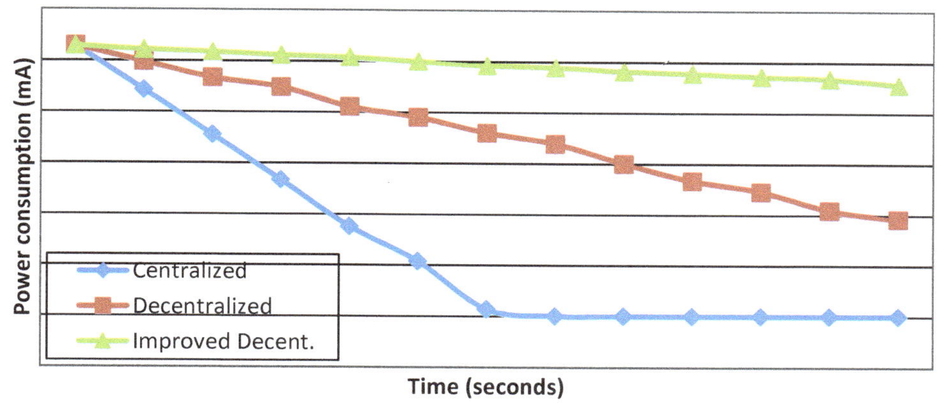

Figure 8. Evaluating the battery life for the reference nodes in the three systems

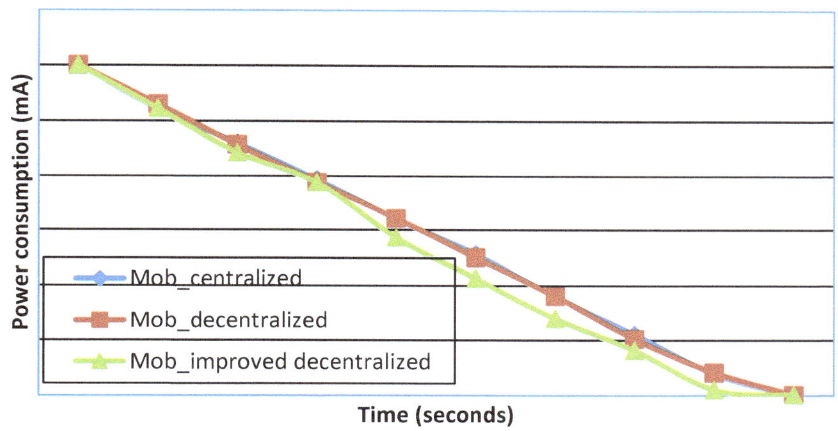

Figure 9. Evaluating the battery life for the mobile target nodes in the three systems

6. Discussion

In WSN, energy has three main consumers: signal processing, data transmission, and hardware operations. In this research work, we aim to reduce the power consumption for both signal processing and data transmission. In the improved decentralized approach, reference nodes are not required to process the mobile targets' locations. Moreover, the total number of messages transmitted over the network is reduced.

Centralized based localization approaches [16, 17, 18] are much more power consuming than decentralized approaches [21, 22, 24, 25], since in centralized systems, usually the following details are required to be transmitted to a sink node: sensor id, target id, packet number and sensor-to-target distance. Therefore, deploying centralized systems with a large density of sensor nodes requires high power-consumption, as each localization message must be transmitted to a sink node via multi-hop network.

Decentralized tracking solutions are more attractive for large sensor networks including thousands of nodes, as there is no need to process the localization information at a sink node side, since all the computations are processed at each mobile target, and therefore the final coordinates are only requested to be transmitted to a sink node. The proposed decentralized systems in [21, 22, 24, 25] are efficient for tracking a single or a few number of mobile targets. Though, these systems are impractical in terms of power consumption to track a large number of mobile targets, as this requires a huge number of messages to be exchanged between mobile targets and the sink node, and this will increase both the power consumption and messages dropping rate. For both systems (centralized and decentralized), increasing the number of mobile targets will increase the number of transmitted messages to the sink node.

An improved decentralized approach is proposed in this paper in order to shrink the total number of messages transmitted over the network and hence decrease the power-consumption. In this system, reference nodes are not required to have high processing capabilities, since computations are taking place at each mobile target node. Grouping mobile targets which set/placed in the same transmission range will increase the tracking efficiency in terms of power consumption.

The proposed improved decentralized tracking system achieves low power consumption, and is applicable to track several mobile targets simultaneously with the minimum communication cost, in such applications necessitate tracking a large number of mobile targets. The system proposed in this paper can be deployed in different types of applications including: tracking a large number of employees in a workplace, tracking passengers in an airport, and tracking fire-fighters in hazard situations.

7. Paper Summary

One of the critical technical issues which must be addressed in developing sensor networks for object tracking applications is energy conservation. In this paper, a decentralized tracking approach is proposed to track the location of multiple mobile targets with minimum communication cost among sensor nodes. The proposed approach achieves an efficient tracking system in terms of localization and power-consumption by grouping the mobile targets which set in the same transmission range.

The proposed system aims to increase the lifetime of reference nodes by reducing the total amount of messages transmitted over the network. The mobile target itself requires processing its localization and transmitting it to a group leader. Therefore, it reduces the total number of messages transmitted to the sink node, and consequently reduces the total amount of energy consumed in the tracking process. Performance evaluation in terms of communication cost and power-consumption was conducted using NS2 simulator.

In this work, all of the reference nodes were considered as router nodes, which means that reference nodes have to be awake all the times, and this increase the power consumption for such WSN based tracking systems. For future work, we aim to involve the end-device nodes in the tracking process in order to minimize the power consumption, while achieving reasonable localization accuracy.

REFERENCES

[1] Hamdi, M., Boudriga, N., Obaidat, M. S. (2008) " WHOMoVeS: An optimized broadband sensor network for military vehicle tracking", International Journal of Communication Systems, Vol. 21, Issue 3, pp. 277-300, ISSN:1074-5351.

[2] Son, B.; Her, Y.; Kim, J. (2006) "A design and implementation of forest-fires surveillance system based on wireless sensor networks for South Korea Mountains". Int. J. Comput. Sci. Netw. Secur.(IJCSNS) 6, 124–130.

[3] Antoine-Santoni, T., Santucci, J., Gentili, E., De; Silvani, X., Morandini, F. (2009) "Performance of a protected wireless sensor network in a fire". Analysis of fire spread and data transmission. Sens. J. 9, 5878-5893

[4] Lloret, J., Garcia, M., Bri, D. and Sendra. S. (2009) "A Wireless Sensor Network Deployment for Rural and Forest Fire Detection and Verification. Sensors". Vol. 9 Issue: 11. pp. 8722-8747.

[5] Resch, B., Mittlboeck, M., Girardin, F., Britter, R. and Ratti, C. (2009) "Real-time Geo-awareness - Sensor Data Integration for Environmental Monitoring in the City". In Proceedings of the IARIA International Conference on Advanced Geographic Information Systems & Web Services – GEOWS2009, Cancun, Mexico, pp. 92-97.

[6] Abusaimeh, H, and Shuang-Hua Y. (2009) "Reducing the transmission and reception powers in the AODV." *Networking, Sensing and Control, 2009. ICNSC'09. International Conference on.* IEEE.

[7] Abu Taleb, A, Dhiraj K. P, and Taskin K. (2009) "A Technique to Identify and Substitute Faulty Nodes in Wireless Sensor Networks." *Sensor Technologies and Applications, SENSORCOMM'09. Third International Conference on.* IEEE.

[8] Alhmiedat, T. A., and Yang, S. H. (2008) "A ZigBee-based mobile tracking system through wireless sensor networks". *International Journal of Advanced Mechatronic Systems*, *1*(1), 63-70.

[9] Alhmiedat, T. A. and Yang, S., (2009) "Tracking Multiple Mobile Targets Based on ZigBee Standard", *Proceedings of the 35th Annual Conference of the IEEE Industrial Electronics Society,* Porto, Portugal.

[10] Alhmiedat, T. A. and Yang, S., (2008) "A Decentralized Low-Power ZigBee-based Localization Approach", *Proceedings of the 14th International Conference on Automation & Computing*, Brunel University, Uxbridge, UK, pp. 41.

[11] Alhmiedat, T. A., & Yang, S. H. (2011) "An improved energy efficient approach for wsn based tracking applications". *Proceedings of the 5th IADIS Multi Conference on Computer Science and Information Systems*, Rome, Italy.

[12] Amundson, I, and Xenofon K. (2009) "A survey on localization for mobile wireless sensor networks." *Mobile Entity Localization and Tracking in GPS-less Environnments*, 235-254.

[13] Alhmiedat, T. Abu Taleb, A. and Bsoul, M. (2012) "A Study on Threats Detection and Tracking for Military Applications Using WSNs", *International Journal of Computer Applications* 40(15):12-18.

[14] Alhmiedat, T. A. and Yang, S., (2007) "A Survey: Localization and Tracking Mobile Targets Through Wireless Sensor Network", *Proceedings of the PGNet Conference*, Liverpool, UK, pp. 48.

[15] Benavoli, A., and Chisci, L. (2007) "Improved estimation for object localization via sensor networks", *Networking, Sensing and Control, IEEE International Conference*, pp. 302-306,.

[16] Alippi, C., and Vanini, G. (2006) "A RSSI based and calibrated centralized localization technique for Wireless Sensor Networks", in *Proc. of IEEE International Conference on Pervasive Computing and Communications (PerCom)*, Pisa, Italy, pp. 301–305.

[17] Sugano, M., Kawzoe, T., Ohta, Y., and Murata, M. (2006) "Indoor Localization System using RSSI Measurement of Wireless Sensor Network based on ZigBee Standard.

[18] Doherty, L., Pister, K. S. J., and El Ghaoui, L, (2001) “Convex Position Estimation in Wireless Sensor Networks,” *IEEE INFOCOM, vol.3*, pp. 1655-1663.

[19] Mensing, C, and Jimmy, N. (2010) "Centralized cooperative positioning and tracking with realistic communications constraints." *Positioning Navigation and Communication (WPNC), 2010 7th Workshop on*. IEEE.

[20] Zhao, Y, Yuan Y, and Marcel K, (2011) "Comparing centralized Kalman filter schemes for indoor positioning in wireless sensor network", *Indoor Positioning and Indoor Navigation (IPIN), 2011 International Conference on*. IEEE.

[21] Lazos, L., Poovendran, R., and Capkun, S. (2005) “ROPE: robust position estimation in wireless sensor networks,” *Fourth International Symposium on Information Processing in Sensor Networks* (IEEE Cat. No. 05EX1086), pp. 31.

[22] Shang, Y., Fromherz , M. P. J., Ruml, W., and Zhang, Y. (2003) “Loclization from mere connectivity,” *Proceedings of the International Symposium on Mobile Ad Hoc Networking and Computing* (MobiHoc), pp. 201-212.

[23] Sanchez-Matamoros, J. M., JRM-D. Dios, and A. Ollero. (2009) "Cooperative localization and tracking with a camera-based WSN." *Mechatronics, 2009. ICM 2009. IEEE International Conference on* IEEE.

[24] Bulusu, N., Bychkovskiy, V., Estrin, D., and Heidemann, J. (2002) “Scalable, ad hoc deployable rf based localization”, *Grace Hopper Celebration of Women in Computing Conf*, Vancouver, British Columbia, Canada.

[25] He, T., Stankovic, J. A., Huang, C., Abdelzaher, T., and Blum, B. M. (2003) "Range-Free Localization Schemes for Large Scale Sensor Networks," *Proceedings of the Annual International Conference on Mobile Computing and Networking, MOBICOM,* pp. 81-95.

[26] Cakiroglu, A, and Cesim E. (2010) "Fully decentralized and collaborative multilateration primitives for uniquely localizing WSNs." *EURASIP Journal on Wireless Communications and Networking*, Vol. 2010, Article ID: 605658,

Permissions

All chapters in this book were first published in IJWMN, by AIRCC Publishing Corporation; hereby published with permission under the Creative Commons Attribution License or equivalent. Every chapter published in this book has been scrutinized by our experts. Their significance has been extensively debated. The topics covered herein carry significant findings which will fuel the growth of the discipline. They may even be implemented as practical applications or may be referred to as a beginning point for another development.

The contributors of this book come from diverse backgrounds, making this book a truly international effort. This book will bring forth new frontiers with its revolutionizing research information and detailed analysis of the nascent developments around the world.

We would like to thank all the contributing authors for lending their expertise to make the book truly unique. They have played a crucial role in the development of this book. Without their invaluable contributions this book wouldn't have been possible. They have made vital efforts to compile up to date information on the varied aspects of this subject to make this book a valuable addition to the collection of many professionals and students.

This book was conceptualized with the vision of imparting up-to-date information and advanced data in this field. To ensure the same, a matchless editorial board was set up. Every individual on the board went through rigorous rounds of assessment to prove their worth. After which they invested a large part of their time researching and compiling the most relevant data for our readers.

The editorial board has been involved in producing this book since its inception. They have spent rigorous hours researching and exploring the diverse topics which have resulted in the successful publishing of this book. They have passed on their knowledge of decades through this book. To expedite this challenging task, the publisher supported the team at every step. A small team of assistant editors was also appointed to further simplify the editing procedure and attain best results for the readers.

Apart from the editorial board, the designing team has also invested a significant amount of their time in understanding the subject and creating the most relevant covers. They scrutinized every image to scout for the most suitable representation of the subject and create an appropriate cover for the book.

The publishing team has been an ardent support to the editorial, designing and production team. Their endless efforts to recruit the best for this project, has resulted in the accomplishment of this book. They are a veteran in the field of academics and their pool of knowledge is as vast as their experience in printing. Their expertise and guidance has proved useful at every step. Their uncompromising quality standards have made this book an exceptional effort. Their encouragement from time to time has been an inspiration for everyone.

The publisher and the editorial board hope that this book will prove to be a valuable piece of knowledge for researchers, students, practitioners and scholars across the globe.

List of Contributors

Hamza Aldabbas
Software Technology Research Laboratory (STRL), De Montfort University, Leicester, United Kingdom

Tariq Alwada'n
Software Technology Research Laboratory (STRL), De Montfort University, Leicester, United Kingdom

Helge Janicke
Software Technology Research Laboratory (STRL), De Montfort University, Leicester, United Kingdom

Ali Al-Bayatti
Software Technology Research Laboratory (STRL), De Montfort University, Leicester, United Kingdom

NavneetKaur
Department of Electronics and Communication Engineering Lovely Professional University, Phagwara Punjab, India

Lavish Kansal
Department of Electronics and Communication Engineering Lovely Professional University, Phagwara Punjab, India

Mr. Sudhir B. Lande
Asst. Prof. K.I.T.S. Ramtek Dist- Nagpur , India

Dr. J. B. Helonde
Principal I T M Kamthee Dist- Nagpur , India

Dr. Rajesh Pande
Prof. & Vice-Principal R. C. O. E. M. Nagpur, India

S.S.Pathak
Prof. in E&ECE dept. IIT Kharagpur. India

Eyad Taqieddin
Department of Network Engineering and Security, Jordan University of Science and Technology, Irbid, Jordan

Ann Miller
Department of Electrical and Computer Engineering, Missouri University of Science and Technology, Rolla, MO, USA

S. Jagannathan
Department of Electrical and Computer Engineering, Missouri University of Science and Technology, Rolla, MO, USA

Shiqi Xing,
Centre for Real-time Information Networks School of Computing and Communications, Faculty of Engineering and Information
Technology, University of Technology Sydney, Sydney, Australia

Pantha Ghosal
Centre for Real-time Information Networks School of Computing and Communications, Faculty of Engineering and Information
Technology, University of Technology Sydney, Sydney, Australia

Shouman Barua
Centre for Real-time Information Networks School of Computing and Communications, Faculty of Engineering and Information
Technology, University of Technology Sydney, Sydney, Australia

Ramprasad Subramanian
Centre for Real-time Information Networks School of Computing and Communications, Faculty of Engineering and Information
Technology, University of Technology Sydney, Sydney, Australia

Kumbesan Sandrasegaran
Centre for Real-time Information Networks School of Computing and Communications, Faculty of Engineering and Information
Technology, University of Technology Sydney, Sydney, Australia

Suman Dutta
IT Trainer, Logitrain.com.au

Shouman Barua
PhD research student. University of Technology, Sydney

Jishu Sen
MSc research student. University of Technology, Sydney

Leila Sahraoui
Department of Electronics Faculty sciences of Engineering University Baji Mokhtar, Annaba bp 12 el hadjar, Algeria

Djmail Messadeg
Department of Electronics Faculty sciences of Engineering University Baji Mokhtar, Annaba bp 12 el hadjar, Algeria

Nouredinne Doghmane
Department of Electronics Faculty sciences of Engineering University Baji Mokhtar, Annaba bp 12 el hadjar, Algeria

Sunil Joshi
Department of Electronics & Communication, CTE, MPUAT, Udaipur, India

Deepak Gupta
Department of Electronics, V.B. Polytechnic College, Udaipur, INDIA

P.C. Bapna
Department of Electronics & Communication, CTE, MPUAT, Udaipur, India

Neha Kothari
Department of Electronics & Communication, CTE, MPUAT, Udaipur, India

Rashmi Suthar
Department of Electronics & Communication, CTE, MPUAT, Udaipur, India

AMEL BOUFRIOUA
Electronics Department, Technological Sciences Faculty, University Constantine 1, Constantine, Algeria

Mohamed Ababou
National Institute of Posts and Telecommunications - INPT, Rabat, Morocco University of sciences Mohamed 1 -FSO, Oujda, Maroc

Rachid Elkouch
National Institute of Posts and Telecommunications - INPT, Rabat, Morocco University of sciences Mohamed 1 -FSO, Oujda, Maroc

Mostafa Bellafkih
National Institute of Posts and Telecommunications - INPT, Rabat, Morocco University of sciences Mohamed 1 -FSO, Oujda, Maroc

Nabil Ababou
National Institute of Posts and Telecommunications - INPT, Rabat, Morocco University of sciences Mohamed 1 -FSO, Oujda, Maroc

A.A.BOUDHIR
LIST Laboratory, Faculty Of Sciences and Techniques, Tangier Morocco CBAR, Al Yamamah university, Kingdom of Saudi Arabia

M. BOUHORMA
LIST Laboratory, Faculty Of Sciences and Techniques, Tangier Morocco CBAR, Al Yamamah university, Kingdom of Saudi Arabia

M. BEN AHMED
LIST Laboratory, Faculty Of Sciences and Techniques, Tangier Morocco CBAR, Al Yamamah university, Kingdom of Saudi Arabia

ELBRAK SAID
LIST Laboratory, Faculty Of Sciences and Techniques, Tangier Morocco CBAR, Al Yamamah university, Kingdom of Saudi Arabia

Sahabul Alam
Department of Computer Science and Engineering,West Bengal University of Technology, Kolkata,India

Debashis De
Department of Computer Science and Engineering,West Bengal University of Technology, Kolkata,India

Preetha Theresa Joy
Department of Computer Science, Cochin University of Science and Technology, Kochi, Kerala, India

K. Poulose Jacob
Department of Computer Science, Cochin University of Science and Technology, Kochi, Kerala, India

Aniket. A. Gurav
Department of Applied Mathematics Defence Institute of Advance Technology, Pune, India 411 025

Manisha. J. Nene
Department of Applied Mathematics and Computer Engineering Defence Institute of Advance Technology, Pune, India 411 025

Johannes Karlsson
Embedded Systems Lab, Umeå University, Sweden

S.Thirumurugan
Department of Computer Applications, JJ College of Engineering & Technology, Trichirapalli, Tamilnadu, India

E. George Dharma Prakash Raj
Department of Computer Science and Engineering, Bharathidasan University, Trichirapalli, Tamilnadu, India

Jyoti Kolap
Department of Electronics and Telecommunication Mumbai University

Shoba Krishnan
Department of Electronics and Telecommunication Mumbai University

Ninad Shaha
Computer Centre, IIT Bombay

Mourad MELLITI
Department of Computer and Communication Technologies National School of Engineering of Tunis (ENIT)

Salem HASNAOUI
Department of Computer and Communication Technologies National School of Engineering of Tunis (ENIT)

Ridha BOUALLEGUE
6'Tel Research Unit, Higher School of Communications of Tunis, Sup'Com, Tunisia

Tareq Alhmiedat
Department of Computer Science, Zarqa University, Zarqa, Jordan

Amer O. Abu Salem
Department of Computer Information Systems, Zarqa University, Zarqa, Jordan

Anas Abu Taleb
Department of Computer Science, Isra University, Amman, Jordan